OXFORD MONOGRAPHS ON GEOLOGY AND GEOPHYSICS

Series editors

H. Charnock
J.F. Dewey
S. Conway Morris
A. Navrotsky
E.R. Oxburgh
R.A. Price
B.J. Skinner

OXFORD MONOGRAPHS ON GEOLOGY AND GEOPHYSICS

1. DeVerle P. Harris: *Mineral resources appraisal: mineral endowment, resources and potential supply: concepts, methods and cases*
2. J. J. Veevers (ed.): *Phanerozoic earth history of Australia*
3. Yang Zunyi, Cheng Yuqi, and Wang Hongzhen (eds.): *The geology of China*
4. Lin-gun Liu and William A. Bassett: *Elements, oxides and silicates: high pressure phases with implications for the earth's interior*
5. Antoni Hoffman and Matthew H. Nitecki (eds.): *Problematic fossil taxa*
6. S. Mahmood Naqvi and John J. W. Rogers: *Precambrian geology of India*
7. Chih-Pei Chang and T. N. Krishnamurti (eds.): *Monsoon meteorology*
8. Zvi Ben-Avraham (ed.): *The evolution of the Pacific Ocean margins*
9. Ian McDougall and T. Mark Harrison: *Geochronology and thermochronology by the $^{40}Ar/^{39}Ar$ method*
10. Walter C. Sweet: *The Conodonta: morphology, taxonomy, paleoecology, and evolutionary history of a long-extinct animal phylum*
11. H. J. Melosh: *Impact cratering: a geologic process*
12. J. W. Cowie and M. D. Brasier (eds.): *The Precambrian–Cambrian boundary*
13. C. S. Hutchinson: *Geological evolution of south-east Asia*
14. Anthony J. Naldrett: *Magmatic sulfide deposits*
15. D. R. Prothero and R. M. Schoch (eds.): *The evolution of perissodactyls*
16. M. Menzies (ed.): *Continental mantle*
17. R. J. Tingey (ed.): *Geology of the Antarctic*
18. Thomas J. Crowley and Gerald R. North: *Paleoclimatology*
19. Gregory J. Retallack: *Miocene paleosols and ape habitats of Pakistan and Kenya*
20. Kuo-Nan Liou: *Radiation and cloud processes in the atmosphere· theory, observation, and modeling*
21. Brian Bayly: *Chemical change in deforming materials*
22. A. K. Gibbs and C. N. Barron: *The geology of the Guiana Shield*
23. Peter J. Ortoleva: *Geochemical self-organization*
24. Robert G. Coleman: *Geologic evolution of the Red Sea*
25. Richard W. Spinrad, Kendall L. Carder, and Mary Jane Perry: *Ocean Optics*
26. Clinton M. Case: *Physical principles of flow in unsaturated porous media*
27. Eric B. Kraus and Joost A. Businger: *Atmosphere-ocean interaction*, second edition

GEOLOGIC EVOLUTION OF THE RED SEA

Robert G. Coleman
Stanford University

OXFORD UNIVERSITY PRESS   New York
CLARENDON PRESS   Oxford
1993

Oxford University Press

Oxford   New York   Toronto
Delhi   Bombay   Calcutta   Madras   Karachi
Kuala Lumpur   Singapore   Hong Kong   Tokyo
Nairobi   Dar es Salaam   Cape Town
Melbourne   Auckland

and Associated companies in
Berlin   Ibadan

Published by Oxford University Press, Inc.,
200 Madison Avenue, New York, New York 10016

Library of Congress Cataloging-in-Publication Data
Coleman, Robert Griffin, 1923–
Geologic evolution of the Red Sea / Robert G. Coleman.
p.   cm.   (Oxford monographs on geology and geophysics ; no. 24)
Includes bibliographical references and index
ISBN 0-19-507048-8
1. Geology–Red Sea Region.   I. Title.   II. Series.
QE350.52.R43C65   1993
551.46'08'09533–dc20   93-14015

9 8 7 6 5 4 3 2 1
Printed in the United States of America
on acid-free paper

Preface

An attempt to produce a book on the geology of the Red Sea is an audacious act, particularly if it is done by a single author. On the other hand, a multi-authored book would probably entail too many authors whose specialized knowledge would not overlap, and would no doubt represent divergent philosophies as well as some national bias. The intent of this book is to provide an integrated starting point for scientists interested in the geology of the Red Sea for comparative purposes or for specialists who need a comprehensive survey of what has already been accomplished. I can truthfully say, not that all of my biases or national prejudices have been removed, but that I have done my best to avoid such cliches.

There has never been an international geologic group or coalition of countries that has set as its goal the conduction of research leading to understanding of the geology of the Red Sea as a whole. Perhaps the main reason for this lack is that there are eight countries whose shores border the Red Sea: Djibouti, Egypt, Ethiopia, Israel, Jordan, Saudi Arabia, Sudan, and Yemen (Fig. I.1). The great divergence in the political organizations and comparative wealth of these countries precludes a joint scientific effort to understand the Red Sea basin and its surroundings.

One could pose the question: what is so important about the geology of the Red Sea? The evolution of the Red Sea is linked to the plate tectonic global evolution and marks the beginning of the separation of the Arabian plate from the African plate. Thus, the Red Sea area is a natural laboratory where it is possible to study the birth of an ocean from continental rifting to ocean floor spreading. The presence of metalliferous brines in the deeps of the Red Sea demonstrates present-day hydrothermal processes that yield economic deposits of potentially great value. Petroleum production in the Gulf of Suez is proof that continental rifting processes can produce economically viable concentrations of oil. The impressive current and past volcanic activity that has produced new ocean crust in the bottom of the Red Sea axial zone and its southern on-land eruptive action in the Afar depression, as well as widespread volcanic eruptions on the western part of the Arabian plate, provides another geologic phenomenon related to continental rifting. The dramatic variation within the Red Sea basin is revealed in sediments that were initially deposited from freshwater lakes to an evaporite basin and finally from open ocean waters. The sere desert climate surrounding the Red Sea has lain bare the rocks involved in the birth and growth of the basin, an unparalleled environment in which to carry out geological field studies. The evolution of the basin is still so young (about 25 Ma) that large-scale features in the crust and mantle related to rifting and drifting can still be measured indirectly by geophysical methods. Is it any wonder those earth scientists from all over the world continue to study one of nature's great experiments?

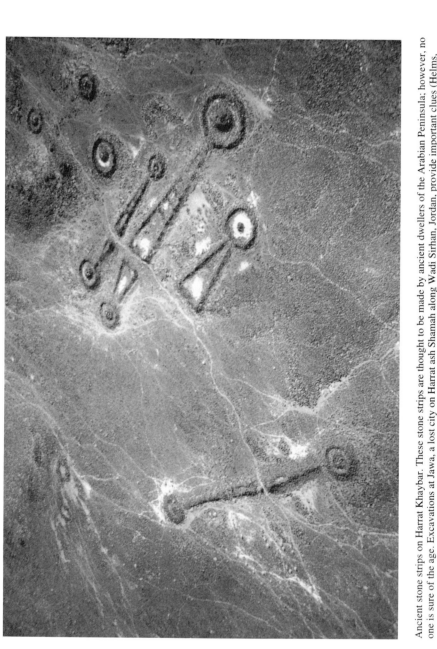

Ancient stone strips on Harrat Khaybar. These stone strips are thought to be made by ancient dwellers of the Arabian Peninsula; however, no one is sure of the age. Excavations at Jawa, a lost city on Harrat ash Shamah along Wadi Sirhan, Jordan, provide important clues (Helms, 1981), indicating a short lived civilization that existed between the 5th and 4th millenia (about 7000 years ago). The ruins are stratigraphically above the stone strips; therefore, it is possible that basalt flows with stone strips on their surface are probably older than 7000 years. The young basalt flows, when they cover the strips, are considered to have occurred between 500 BC and 700 AD, at the same time as the historic Chada flow near Al Medina.

Acknowledgments

Dr. David Ross of Woods Hole Oceanographic Institute invited me to participate in the 1972 Glomar Challenger Deep Sea Drilling Leg 23B, sparking my initial interest in the Red Sea. Many of his ideas have been incorporated into this book. Dr. Glen Brown of the United States Geological Survey (U.S.G.S.) produced an extremely helpful review of the book. He introduced me to the geology of the area in 1969 and over the years his expert knowledge has been a constant source of information to me. For other reviews I would like to thank Andrew Griscom of the U.S.G.S. for his comments on the geophysics section and Professor Ziad Beydoun for his comments and suggestions on the stratigraphy and petroleum potential sections. Dr. Lev Zonenshain provided me with photos and reprints of the *Pisces* submersible (1979–1980) expedition in the Red Sea axial trough. I would also like to acknowledge permission by the Ministry of Petroleum and Mineral Resources, Deputy Ministry of Mineral Resources, Jiddah, Kingdom of Saudi Arabia, and the Director of the U.S.G.S. to use information developed from my research in Saudi Arabia.

There are of course many other earth scientists with whom I have had important exchanges and I would like to thank all of them collectively for their scientific stimulus, and numerous reprints received. I dedicate this book to my wife Cathryn whose constant support was the main incentive for my successful completion of the book.

Contents

GEOLOGIC EVOLUTION OF THE RED SEA

Introduction

Historical Aspects of Geologic Studies in the Red Sea Basin

As in nearly all countries of the Middle East (Fig. I.1), geologic mapping related to mineral and energy exploration is the main incentive for geologic studies in the Red Sea Basin. The most ambitious project to date, initiated to produce geologic maps of the Arabian Peninsula, was a cooperative venture between the Kingdom of Saudi Arabia and the U.S. Government. A total of 21 maps at a scale of 1:500,000 was agreed upon for Saudi Arabia alone, but as the work continued, the

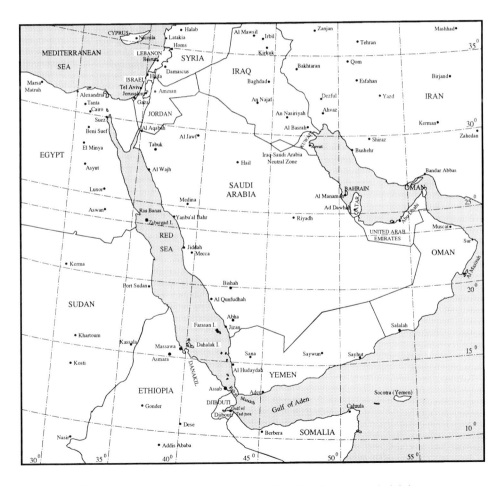

Fig. I.1. Geographic place name map of the Red Sea area and vicinity.

3

need for a smaller scale map at a scale 1:2,000,000 to include all of the Arabian Peninsula also became clear. This map, covering a much larger area, was developed by recruiting geologic map information from 11 separate groups, including adjacent countries and oil exploration companies. Published in 1963 by the U.S. Geological Survey, it was a catalyst, bringing new efforts to the study of the Red Sea (U.S.G.S., 1963). Explanatory volumes accompanying the map provided a new basic stratigraphy and structural framework for future study of the area (U.S.G.S., 1963; Geukens, 1966; Greenwood and Bleackley, 1967; Bender, 1975; Brown et al., 1989). The economic advantage produced by petroleum production brought further support for the complete mapping of the western part of Saudi Arabia. The geologic studies were multinational and involved geologists from France (BRGM), Japan, Great Britain, Germany, Austria and the United States. A tectonic map (1:2,000,000) of the Arabian Peninsula (Brown, 1972) consolidated geologic information all along the eastern Red Sea coastline as well as showing bathymetry of the axial trough. In 1980 the Israel Geological Survey published the Sinai Geological map (1:500,000) (Eyal et al., 1980) the research for which was carried out during the Sinai occupation from 1967 to 1984. A later landsat map (1:500,000) of the Sinai has been published by the French (Henry and Chorowicz, 1986). The publication of the geologic map (1:500,000) of the Yemen Arab Republic (Grolier and Overstreet, 1978), along with the other maps listed above, made geologic maps at the scale 1:500,000 available for the entire eastern margin of the Red Sea. A short history of this activity between 1963 and 1984 can be found in Brown et al. (1989).

The Red Sea coastal plain of Egypt is well known and has been mapped at various scales (Fig. I.1). The recent monograph *The Geology of Egypt* (Said, 1990) provides an excellent summary of the geologic research in the area. A new map of Egypt (1:500,000), now available as part of the monograph by Said (1990), incorporates earlier geologic mapping. The coastal areas of the Sudan have received the least geologic mapping and research of all the countries bordering the sea. New maps (Vail, 1978; Roberston, 1988) update the older map produced by the Sudanese government (Sudan, 1963), and several books on the geology of Sudan (Whiteman, 1971; Vail, 1988) are available. Petroleum exploration has led to the production of some reports on the Red Sea Basin sediments of the Sudan (Carella and Scarpa, 1962; Sestini, 1965; Bunter and Abdel Magrid, 1989a,b; Montenant et al., 1990).

The Ethiopian Red Sea coastal plain (Fig. I.1) is quite well known as a result of petroleum exploration and intensive scientific studies centered in the Afar Depression. A geologic map of Ethiopia (1:2,000,000) compiled by Kazmin (1973) is the best overall source for regional geology. Contemporaneous with its publication is the 1:500,000 geologic map of northern Afar that was produced by the C.N.R.S. (France) and C.N.R. (Italy) expeditions (Barberi et al., 1971). Later a map (1:500,000) of central and south Afar was produced (C.N.R., 1975). This multinational study produced a synthesis of the volcanic, sedimentary, and tectonic characteristics of the area, combined with geophysical observations. Earlier larger scale maps of the Danakil Depression had been published (Brinckmann and Kursten, 1969). A map (1:2,000,000) of Djibouti, which faces the straits of Bab El

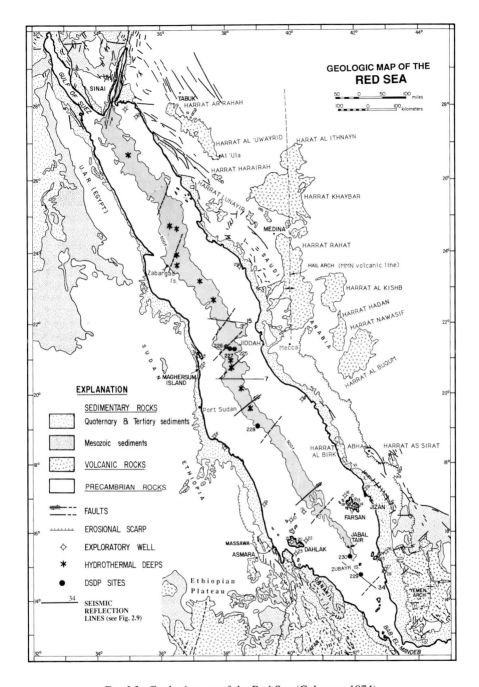

FIG. I.2. Geologic map of the Red Sea (Coleman, 1974).

5

Mandeb and is the most southern territory on the western Red Sea coast, was also published (Clin and Pouchan, 1970). A map of Somalia and Ethiopia at the scale of 1:2,000,000 has been published by C.N.R. Italy (Merla, 1979), providing nearly complete geologic mapping of the western margin of the Red Sea.

As part of the *Glomar Challenger* deep sea drilling project (DSDP) report on the Red Sea Leg 23B, I compiled a geologic map of the Red Sea (1:2,000,000; Coleman, 1974) (Fig. I.2). Some 20 years later, this map has not been updated, nor have further compilations been made. During the period when I first started work in the Red Sea area the arrival of satellite imaging revolutionized the development of geologic maps. The accuracy of location and lack of distortion on satellite images allows them to be used as bases for geologic maps, and enhancement of the images brings out structural features not easily observed by ground observations. Today many of the current geologic mapping projects in the Red Sea Basin are based on satellite images. Examples are the geologic maps of the Sinai (Eyal et al., 1980) and Yemen (Grolier and Overstreet, 1978), where the great expense of producing topographic maps for geologic mapping was avoided through the use of these new techniques.

Historically the Red Sea has been an important seaway and has been chronicled in numerous ways. The most useful and succinct reference to the older exploration and research is given by Edwards and Herad (1987). In 1869, the opening of the Suez Canal transformed the Red Sea into the most important link between Europe and the Far East. This new easy access to the Red Sea brought many research ships to the area, and observations were also made in transit by expeditions heading for the Indian Ocean (Backer, 1975; Edwards and Herad, 1987), during the late 19th and early 20th centuries.

The continuing improvement of acoustic surveys of ocean bottom sediments, as well as accurate measurements of gravity, magnetics, and heat flow, led to new discoveries within the Red Sea (Falcon et al., 1969; Backer, 1975). In 1966 the discovery of hot brine pools stimulated several expeditions, leading to the publication of *Hot brines and recent heavy metal deposits in the Red Sea* (Degens and Ross, 1969). During the same year a symposium brought together scientists from several disciplines to discuss the structure and evolution of the Red Sea, and the nature of the Red Sea–Gulf of Aden–Ethiopian rift junction (Falcon et al., 1969). This effort was a strong catalyst in bringing about collaborative studies, as the new and evolving ideas of plate tectonics were shown to be important in explaining the Red Sea evolution (McKenzie et al., 1970). In fact, the meeting can be taken as the starting point of two decades of cooperative research that still continues. On April 12, 1972, the R/V *Glomar Challenger* entered the Red Sea and drilled at six sites in and around the axial rift (Whitmarsh et al., 1974). These drillings confirmed that the axial trough was an active spreading center, giving rise to hot heavy metal brines. However, because the drilling did not penetrate the thick evaporite sequence along the edge of the axial trough, the nature of the underlying basement remained enigmatic.

In 1974, a symposium held in Bad Bergzaben (Pilger and Rosler, 1974a,b) brought together a large group of European geophysicists, geologists and volcanologists, many of whom had initiated field work in the Afar Depression in the late 1960s. At the time of this meeting the plate tectonic theory was gaining momentum and the Afar Depression was considered to be a triple junction, where the Red Sea, East African, and Gulf of Aden rifts terminated. Spectacular volcanic activity in this region confirmed the idea of new crust formed by upwelling mantle magma, and geophysical measurements of the lithosphere illustrated attenuation and uprise of the asthenosphere.

In 1977, the Saudi Arabian government published the results of the *Chain, Glomar Challenger,* and *Valdivia* expeditions in the Red Sea (Hilpert, 1977). These papers consolidated the Red Sea oceanographic studies, up to that time, and provided a forum for new ideas on Red Sea geology. In April 1979, a multidisciplinary international meeting in Rome, the "Geodynamic evolution of the Afro-Arabian rift system" focused on rifts from Kenya to Jordan and developed new insights, particularly on volcanism in Yemen, Saudi Arabia, and Jordan as related to the Red Sea opening. These papers discussed apparent plate tectonic relationships between mantle melting using petrologic and geochemical parameters rather than extensive geophysical measurements (Zanettin, 1980).

One of the vexing problems in understanding the Red Sea development has been trying to characterize the crust underlying the marginal trough and coastal plain. In 1979, the U.S. Geological Survey and the Directorate General of Minerals, Saudi Arabia carried out a deep seismic refraction profile across western Saudi Arabia and into the southeastern Red Sea (Healy et al., 1982). A workshop in Park City, Utah, USA (August 1980) brought together an international group of 46 seismologists who had previously been given the refraction data sets derived from the Saudi refraction line. Numerous interpretations were presented, some of which were later published (Mooney and Prodehl, 1984; Mooney et al., 1985). The origin of the transition zone between the Arabian shield and the Red Sea was an area of much concern, but the new data clearly showed a dramatic change from a 40 km continental crust to oceanic crust of only 17 km. This new data brought to a close the earlier idea that the Red Sea was a large single graben underlain by only continental crust (Cloos, 1930).

By the mid 1980s the growing concern about environmental changes within the world's oceans led to the designation of the Red Sea as a *key environment* (Edwards and Herad, 1987). Because of its highly diverse and unique environment, and its great scientific and ecological importance (Ross, 1983) a special volume was published in 1987. *Key environments–Red Sea* contains important facts concerning the human, geologic, and biologic history of the Red Sea Basin. In this volume geologists, biologists, archeologists, and ecologists define the Red Sea as a key environment, one important enough to initiate conservation measures to preserve it (Edwards and Herad, 1987). Also published in the same year, *Afro-Arabian geology* (Bowen and Jux, 1987), a book resulting from an international conference (Bowen, 1984a,b), provides a broad background for the region surrounding the Red Sea.

The special issue of *Tectonophysics* on Zabargad Island (Bonatti, 1988) is a collection of papers dealing mainly with the unique mantle-basement rocks exposed on the western side of the Red Sea near Ras Banas. High temperature and high pressure mineral assemblages preserved in the complex support the idea that the Zabargad complex is similar to "core complexes" exposed by extensional faulting during attenuation of the continental crust.

Exploration for oil in the northern Red Sea has produced important analyses of the Red Sea sediments. These reports were published by the French oil company TOTAL (Berthon et al., 1987) and in a special issue of the *Bulletin of the French Geological Society* on the Gulf of Suez and Red Sea (Purser et al., 1990a, b). In 1986 a workshop, the "Gulf of Suez and Red Sea Rifting," was convened in Hurgada, Egypt (Le Pichon and Cochran, 1988). Nearly all the papers favored characterizing the Suez as a failed extensional rift with the stretching terminated by the Late Miocene Dead Sea Rift. Discussions of stratigraphy and structure at this workshop incorporated much of the older data in the context of models that are compatible with extension and known heat flow measurements. A complementary special issue of *Tectonophysics* on the Elat (Aqaba)–Dead Sea–Jordan subgraben system, provides new and important information on the structure and relationship of this area to the Red Sea (Ben-Avraham, 1987).

A special issue of the *Journal of African Earth Sciences* on African rifting (Kogbe, 1989b) includes several papers on the structural aspects of rifting in the Red Sea and its relationship to African rift systems. A special volume of *Tectonophysics* resulting from the symposium "World rift systems" held at Texas A&M University in 1989 includes five papers dealing with structural and igneous history of the Africa/Red Sea area (Gangi, 1989).

In 1991 a special issue of *Tectonophysics* consolidated a large amount of geophysical data and provides new ideas on the origin of the Red Sea (Makris et al., 1991b). The following year another special issue of the same journal on "The Afro-Arabian rift system" contained a mixture of papers on geophysics, tectonics, and petrology related to rifting (Altherr, 1992).

In 1986, with the support of the World Bank, the Red Sea and Gulf of Aden Regional Hydrocarbon Assessment (Study) was launched in response to the lowering of oil prices and the desire to develop a multinational Red Sea Basin exploration program. This study developed a strategy which involved the collecting, integrating, and common archiving of Red Sea Basin-wide data-sets, and focused on technical analyses and evaluation of areas with economic interest. These results, summarized in two special issues of the *Journal of Petroleum Geology* (Beydoun, 1989; Beydoun and Sikander, 1992a, b), provide much new information on the Red Sea Basin and its hydrocarbon potential.

The past 20 years surely have been a period of exciting research in and around the Red Sea, one in which studies have confirmed a history closely tied to recent global plate movements. In addition to the symposia and special issues of journals on the Red Sea, many books and papers have been published over the past 20 years. Some have had a profound influence on the thinking about the Red Sea, while others are better left in obscurity.

My own experience with the geology of the Red Sea Basin started in 1970 when I managed to collect samples from the striking Jabal Tirf igneous complex. A few years later I carried out mapping of the Precambrian rocks in the Khamis Mushayt area. This work was followed by a joint U.S.G.S. reconnaissance study of the coastal plain rocks from Ad Darb to the Yemen border along the eastern shore of the Red Sea. A regional study in the early 1980s of all the volcanic harrats in western Saudi Arabia convinced me that they must be part of the Red Sea opening. At the same time as my Saudi Arabian studies took place, I was also able to do research in the Sultanate of Oman, research that provided me with a regional perspective of plate tectonics of the Arabian Peninsula, and as the designated petrologist aboard the *Glomar Challenger* Leg 23B (1971), I had the chance to study some of the first basaltic rocks recovered from the active spreading center of the Red Sea.

The material that follows has been sifted carefully. The intent is to present an introduction to the geology of the Red Sea. Current and future studies will continue to change these interpretations, but the basic geologic data presented in this book should serve as a departure for new models and explanations for future studies.

1

Geomorphology

The Red Sea area is dominated by an arid climate, and nearly all of the erosional processes are directly related to these desert conditions. Wind erosion plays a dominant role in the development of the present landscape. Torrential rains are scarce but provide the only source of water for rapid erosion (Fig. 1.1); there are no permanent rivers that flow into the Red Sea at present. Much of the spectacular relief in the area is related to tectonic movements resulting from the opening of the Red Sea, as explained below. There have been no systematic geomorphic studies of the Red Sea Basin; however, Brown et al. (1989) give a very detailed account of geomorphology in their paper "Shield area of western Saudi Arabia, Geology of the Arabian Peninsula."

Shape of the Red Sea Basin

The Red Sea forms an elongate depression over 200 km long. In the northern part, above 24°N, the subparallel shorelines are fewer than 180 km apart; southward the shorelines become irregular, with separations of more than 350 km (Fig. 1.1). The southern part of the Red Sea narrows to about 28 km at the Straits of Bab el Mandeb where it connects with the Gulf of Aden. The bathymetric chart of the Sea (Laughton, 1970) reveals a distinct main trough extending northward from the Zubayr Islands to the southern tip of the Sinai Peninsula. This trough is cut by an axial trough 5 to 30 km in width (Fig. 1.1). Selected topographic cross-sections of the axial trough reveal steep-sided walls and very irregular bottom topography. Segments of the axial trough are marked by changes in their elongate directions, perhaps as a result of contemporaneous faulting. High-temperature brine pools occupy the deepest parts of the axial trough. Tombolos with volcanic centers along the shorelines indicate rapid basin deposition combined with isolated volcanic centers within the axial zone (Fig. 1.2).

The shelf segments within the Red Sea are shallow and flat. South of 21°N the shelf merges imperceptibly with the coastal plain; however, north of this latitude the shelf becomes narrower and is interrupted by rather sharp topographic breaks. Onshore, the coastal plains broaden south of 21°N, reaching a width greater than 50 km (Fig. 1.3) and in restricted areas may be covered by recent lava flows. North of this latitude, the coastal plains are narrower and contain raised terraces that represent older shorelines (Brown et al., 1989). These features indicate emergence (or uplift) in the north and submergence combined with rapid deposition in the south.

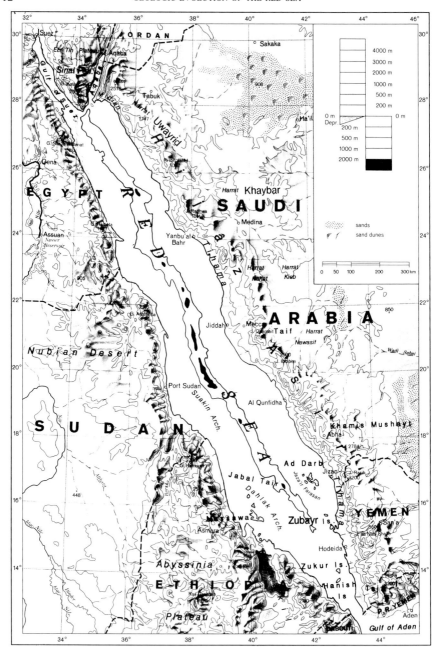

Fig. 1.1. Topographic map of the Red Sea prepared by Spohner and Oleman (1986).

Landward from the littoral zone and coastal plain the topography is quite variable. On the Saudi Arabian side a magnificent erosional escarpment extends unbroken southward from Taif to Taizz, with steep slopes facing the Red Sea, and slopes gently eastward away from the Basin. The crest of the scarp has been

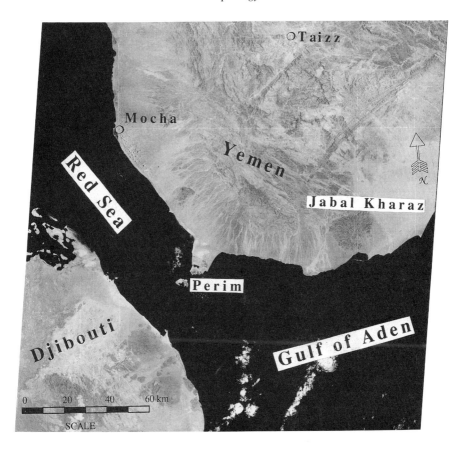

Fɪɢ. 1.2. Southern end of the Red Sea. Perim Island and Turba peninsula consists of recent volcanics nearly closing the straits of Bab el Mandeb. Taizz is centered on the dissected Yemen volcanic plateau with elevations in excess of 3000 m. Djibouti is covered by both recent and Miocene–Pliocene mafic lavas. Both shorelines are actively aggrading to produce narrow alluvial coastal plain covering the lavas. Landsat-I Image E-1045-06560-7, Sept. 6, 1972.

deeply eroded, forming an irregular scalloped pattern that is subparallel to the Red Sea axis. These scarp mountains average 2000 m south of 21° N and reach a maximum elevation of more than 3000 m at Jabal as Sudah in Saudi Arabia and at Jabal as nabi Shaib in Yemen (Fig.1.4). To the north, the scarp mountains rarely rise above 1000 m but extend across the Gulf of Aqaba, forming the eastern boundary of the Gulf of Suez. Modern geologic maps of Saudi Arabia (Brown et al., 1989) do not show a single master normal fault that is responsible for this scarp; instead numerous older faults are shown intersecting the scarp at high angles. In contrast, major subparallel faults mark the Ethiopian scarp. Normal faulting is responsible for this scarp (Mohr, 1971).

At Massawa in Ethiopia the scarp turns southward, and leaves the Red Sea trend and connects with the East African rift system (Figs.1.1 and 1.5). The scarp

Fig. 1.3. Western shoreline of the Red Sea along the Ethiopian coastline. The narrow alluviated coastal plain has fringing coral reefs. Dissected terrain west of the coastal plain is uplifted Precambrian basement, which makes up the northeastern margin of the Ethiopian Plateau, with elevations in excess of 2000 m. Landsat-I Image E-1103-07182-7, Nov. 3. 1972.

forms the western boundary of the Danakil (Afar triangle) Depression. Here extensive Neogene volcanism and extensional block faulting have produced a complex mixture of interior basins and volcanic landscapes (Tazieff et al., 1972). East of the Danakil Depression the southwestern boundary of the Red Sea consists of the Danakil Alps, which form an elongate ridge trending southward to the Gulf of Tadjura. The triple junction between the Red Sea–Gulf of Aden–East African rift systems is located within the Afar triangle (Barberi et al., 1972). The topography is marked by numerous volcanic vents and flows dislocated by normal faulting, which develops with the continued extension of the region (Marinelli et al., 1980). The Red Sea coastal plains in Sudan and southern Egypt are narrow, bordered by uplifted Precambrian basement, the lower slopes of which merge seaward and consists of interdigitating fans and pediments pierced by outliers of the basement (Whiteman, 1968; 1971).

Fɪɢ. 1.4. Crest of the scarp just south of Abha. The nearly level surface has an elevation of about 2000 m and slopes gently eastward. The major uplift began 13.5 Ma ago according to fission track ages (Bohannon et al., 1989). View from helicopter looking due south (1971).

The northern terminus of the Red Sea is split by the Gulf of Suez and Gulf of Aqaba (Elat). The Gulf of Aqaba is about 180 km long and 25 km wide, narrow in the north and widening to the south. It has a maximum depth of 1850 m, with the surrounding mountains rising to 1.5 km. There are large alluvial fans along the west side of the Gulf, but no shelves or coastal plains. It is considered to be a fault-controlled depression related to the Dead Sea fault and is partly filled with clastics (Garfunkel et al., 1976; Ben-Avraham et al., 1979a) (Fig. 1.1).

In contrast, the Gulf of Suez is a northward extension (300 km) of the Red Sea, separating the African Plate from the Sinai Plate. The widest part does not exceed 50 km and, unlike the Gulf of Aqaba, the water is not more than 100 m deep. The Gulf of Suez is considered to be a half-graben controlled by normal listric faulting that has formed a flat-bottomed shallow trough. There has been no recent volcanic activity within or surrounding the the Gulfs of Suez and Aqaba (Abel-Gawad, 1970).

Red Sea Islands

Three distinct kinds of islands are present within the Red Sea, discounting the many smaller barrier islands related to coral reef formation. The volcanic islands of the southern Red Sea (Jabal Tair, Zubayr Group, Zukur Group, and the Great

Fig. 1.5. Western side of the Red Sea illustrating the scalloped nature of the Dahlak Archipelago shoreline, controlled by salt dome dissolution. The extreme northern end of the Danakil Depression (Afar) terminates in the Gulf of Zula. Dark material in the Danakil represent recent volcanic mafic eruptions. The northern extension of the Danakil Horst can be seen in the lower left corner. Landsat-I Image E-1102-07130-7, Nov. 2, 1972.

and Little Hanish) consist mainly of pyroclastic cones and interlayered flows (Gass et al., 1973). They are found within the axial part of the Red Sea. Jabal Tair, the most northerly, is considered still active, forming a small shield volcano approximately 4 km in diameter. Evidence of recent flows can be seen on the surface, and the effects of erosion are not apparent. The Zubayr Island group is made up of intermediate basalt and the most southerly Hanish–Zukur Island group is mainly alkali basalt with some silicic differentiates. These more southerly volcanic islands are no longer active and have undergone fairly strong erosion (Gass et al., 1973). As pointed out earlier, the many small circular-shaped topographic highs shown on Laughton's map (Laughton, 1970) probably represent individual submarine volcanoes (Coleman, 1974). The deep submersible mapping carried out by Russian scientists revealed that the axial zone is replete

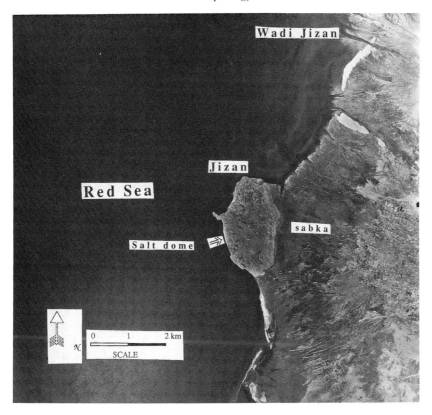

Fig. 1.6. The Jizan salt dome forms a circular dome bisecting the eastern shoreline of the Red Sea, rising 20 to 40 m above the coastal plain. Doming continues and has diverted active wadis (streams) to the north, where active alluvial deposition continues. The city of Jizan is situated on the north end of the salt dome, and the dark areas surrounding the dome are sabkahs. Aerial photo GSMO 8 129, Jan. 22, 1951.

with many small domal eruptives (Monin et al., 1982). It seems quite likely, then, that in some instances the slow spreading nature of the Red Sea axial zone could lead to the formation of many individual small volcanic domes or larger submarine peaks.

The Dahlak and Farasan archipelagoes form symmetrical low-lying island groups on opposite sides of the southern Red Sea. These islands consist of marine sediments, evaporites, and coral reef limestone not exceeding 20 m above sea level. Exploratory oil wells have penetrated Miocene salt at various depths; the islands, and the peculiar scallop-shaped shorelines, provide evidence that salt domes reached the surface. On the Saudi Arabian coast at Jizan a small salt dome rises above the coastal plain (Fig. 1.5). Within the Farasan and Dahlak Islands, the salt has been dissolved by the seawater, and producing the unique, scalloped shorelines that clearly delimit the position of the salt dome (MacFadyen, 1930; Frazier, 1970) (see Fig. 1.1).

Zabargad Island represents the third kind of island present in the Red Sea. It consists of both mantle and lower crustal metamorphic rocks. This small island (about 4.5 km²), located on the western shelf of the Red Sea (23°37' N - 36°12' E), is triangular in shape, rising about 235 m above sea level. It is made up of peridotites, and amphibolites intruded by diabase and is covered by Tertiary marine sediments. Recent interpretations have suggested that the island represents a "core complex" exposed during extensional faulting related to the Red Sea opening (Bonatti et al., 1983).

Coral Reefs

The development of coral reefs in the Red Sea has been continuous since its inception, and today coral reefs are present everywhere in the Sea except in the upper reaches of the Gulf of Suez. The best-developed reefs are found in the central and northern areas. The most common reef type is the fringing reef. Along the steep shores of the Gulf of Aqaba, the reefs cling to bare Precambrian basement rock and the canopy grows outward. In the Sudan, the seaward edge of a reef may be more than 1 km from the shore, with a significant lagoon in between (Head, 1987). Barrier reefs in the Red Sea are thought to form as part of active faulting along shorelines (Braithwaite, 1987). The fringing reefs of the Red Sea are often penetrated at intervals by narrow channels. These T-shaped openings, which are called *sharms* in Saudi Arabia and *marsas* on the Sudan side of the Red Sea (Head, 1987), are thought to be channels formed by erosion in the pluvial Pleistocene and drowned by post-glacial sea level rise (Brown et al., 1989). During the last glacial period, sea level fell by at least 120 m (Braithwaite, 1987), and the present day seasonal streams (wadis) are often still connected to these earlier erosional features. Brown (1989) has estimated that the streams formed about 12,000 to 8,000 years ago during the early part of the Pluvial period.

Axial Trough

Because it is the locus of formation of new oceanic crust and active normal faulting (Figs. I.1 and I.2) the axial trough has been studied in great detail over the past twenty years. Before 1980, no observations of the trough had been made by submersible craft, and all observations of its morphology were by indirect sounding and dredging (Monin et al., 1982). The axial zone is surrounded by fault terraces that step up to the edge of the axis. The almost vertical faulted faces of the walls may rise 500 to 800 m. The rift axis is about 4 to 5 km wide at depths of about 2000 m and has a very rugged terrain. Stretching through the central part of the zone a string of young basalt volcanoes from 200 to 300 m high compose the extrusive zone. They are composed of lava pipes that extrude magma from the top. The floor of the trough consists mainly of silt interlayered with the more ancient flows. Large open fissures appear in clusters along the extrusive zone, breaking up the silt–lava sequences and often transecting the earlier-formed volcanoes. These fissures may make up 6% of the internal rift, and their parallelism with the main axis is a manifestation of active extension within the zone. The lack of covering sediments on the volcanic and extensional features in the zone attest to its youthfulness (Monin et al, 1982).

2

Stratigraphy

The sedimentary record of the Red Sea Basin provides critical evidence toward establishing a coherent explanation of its formation. The record is not complete because active sea-floor spreading enlarges the axial trough, and clastic sediments from surrounding escarpments provide debris that covers earlier-formed sediments and volcanics along the coastal plains. Stratigraphic evidence presented in this section is derived from reports and geologic maps of the surrounding area, data released from petroleum exploration companies, and oceanographic studies

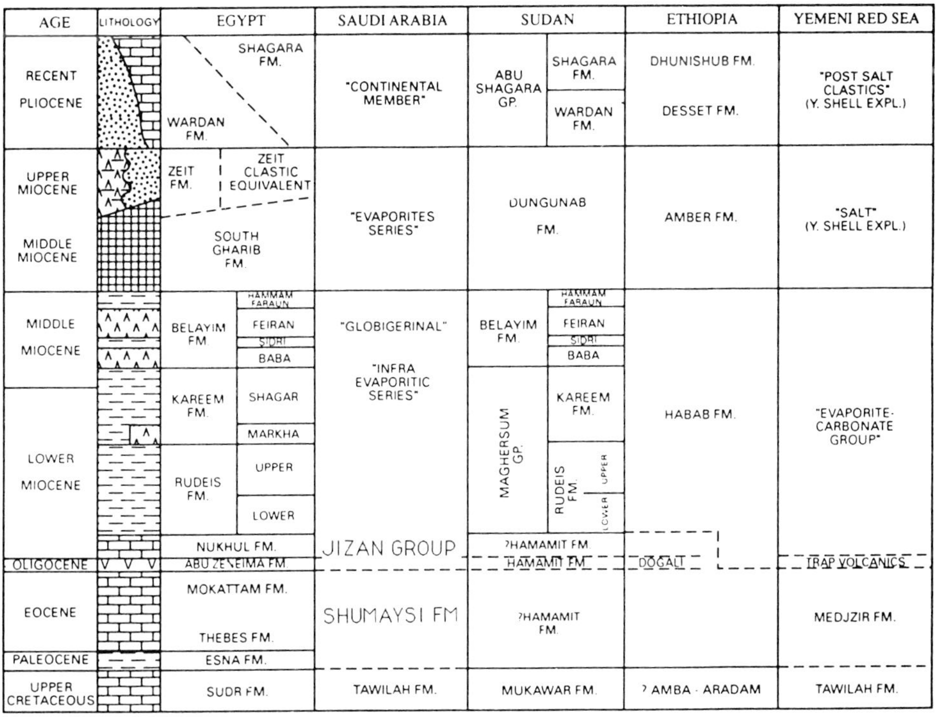

FIG. 2.1. Correlation chart of the Red Sea Basin tertiary sediments from the Gulf of Suez, Sudan, Saudi Arabia, Yemen, and Ethiopia. Nomenclature adapted from the World Bank/United Nations Development Programme Red Sea–Gulf of Aden study (O'Connor, 1992). Modified from Hughes and Beydoun (1992), with permission.

in the Red Sea area. Some of this data was found in reports of a restricted nature, whose contents were difficult to verify. The Red Sea and Gulf of Aden Regional Hydrocarbon Assessment study, sponsored by the World Bank, has produced a new synthesis of the Red Sea Basin stratigraphy, which has been used extensively in this discussion (O'Connor, 1992). The following discussion divides the sedimentary history into five sections: (1) basement sediments, (2) pre-rift sediments, (3) syn-rift sediments, (4) evaporites, and (5) post-evaporite sediments. The lithostratigraphic chart provides a guide for understanding the Red Sea Basin stratigraphy and correlations (Fig. 2.1).

Basement Sediments

Nearly all of the exposed rocks surrounding the Red Sea Basin are Precambrian in age and have undergone repeated deformation, intrusion, and metamorphism. These rocks appear to have been derived from oceanic suites of basaltic and andesitic rocks associated with flysch and deep ocean basin deposits. The mobilized basement and intrusive rocks yield ages of 500 to 900 Ma, representing Late Phanerozoic cratonization (Clifford, 1970; Stacey and Hedge, 1984). In East Africa the Late Precambrian Pan-African event deformed and mobilized

FIG. 2.2. Wajid sandstone resting on Precambrian basement rocks is offset by normal faulting. The downthrown block is rotated, and dipping toward the Red Sea. South side of Jabal Abu Hassan, 100 km northwest of Jizan. View from helicopter facing north (1971).

FIG. 2.3. Wajid sandstone dipping steeply toward the Red Sea and rotated by listric normal faulting. Located on the edge of the Tihama coastal plain north of Ash Shuqayq. View to the south with Red Sea shore to the right, from a helicopter at about 100 m elevation.

this crustal area in north and northeast-trending belts that are distorted and invaded by circular plutons (Greenwood et al., 1976). The Pan-African belt extends through the eastern side of Africa into Arabia. The basement rocks on either side of the Red Sea are remarkably similar, and major structural trends can be matched after palinspastic reconstructions (Greenwood et al., 1976; Vail, 1985). The present trend of the Red Sea axial trough where active spreading is now taking place has tenuous relations with the much earlier Pan-African tectonic trends.

The crystalline basement was certainly stabilized by the end of the Precambrian and in the vicinity of the Red Sea had undergone deformation, metamorphism, and erosion. Included in the basement rocks are quartz sandstone of the Nubian type that were deposited all across the shield during the Early Paleozoic. These sandstones are generally nonfossiliferous, consisting mostly of cross-bedded aeolian and fluvial nearly pure, highly indurated quartz sandstones (Brown, 1970, Brown et al., 1989). This sandstone unit is referred to as the Saq sandstone, and ranges from Cambrian (?) to Early Ordovician in age (Powers et al., 1966). In Arabia the outcrops of the Saq (Wajid) sandstone form arcuate patterns, extending from southern Sinai and Jordan across Saudi Arabia, framing the eastern edge of the shield. At the Yemen–Saudi Arabia border this sandstone is found along the crest of the high scarp, and small patches are also found along the coastal plain at sea level (Hadley and Schmidt, 1980; Bohannon, 1986a), where the sedimentary units have been displaced by normal faulting related to Tertiary extension of the Red Sea (Figs. 2.2 and 2.3). In the Ethiopian highlands, Nubian type sandstones are found above Precambrian basement. The lack of fossils in these sandstones precludes an exact age designation. It is possible that some of these sandstones extended over much of the East African and Saudi Arabian shield areas during most of the Paleozoic and into the Mesozoic (Beyth, 1991). The presence of Upper Ordovician glacial deposits in central Saudi Arabia has been linked to the ice cap that existed at that time on the Gondwana paleocontinent (Vaslet, 1990). These deposits consist of tillite, fluvial glacial, and glacial marine or lacustrine found in the Zarqa and Sarah formations. These deposits rest unconformably on the Qasim formation of Late Ordovician age, but also rest on the Saq (Wajid) sandstones where deep erosion has prevailed (Vaslet, 1990). Reconstructions of the paleogeography using these glacial deposits in West Africa, Turkey, Iran, and Syria indicate that Arabia belonged to the Gondwana paleocontinent at this time (Vaslet, 1990).

The older basement in the central portion of the Red Sea Basin may consist of Precambrian crystalline rocks, or it is also possible that the older Nubian type sandstones may have been confused with some of the Cretaceous Nubian type sandstones. There seems to be no observable stratigraphic record from the early Paleozoic through the beginning of the Mesozoic in the Red Sea Basin (Fig. 2.1). This stratigraphic record suggests that the Red Sea Basin was a positive

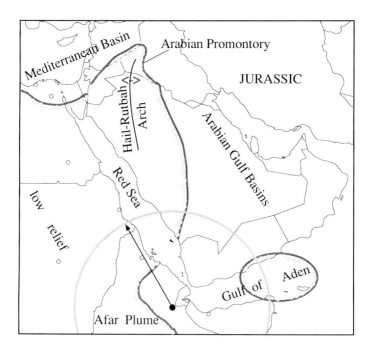

Fig. 2.4. Paleogeographic reconstruction of the Tethyan shoreline for the Jurassic. The dots along the bold line indicate landward areas; the large circle indicates the possible influence of uplift from a hot spot centered in the Afar. No palinspastic reconstruction of plates is given and only the offset along the Dead Sea Rift is shown.

area characterized by minor erosion, glacial scouring and reworking of aeolian sand deposits during this time. To the east, in the Rub Al Khali and Central Arabia, the shelf deposits began to accumulate along the edge of the continent, beginning in the Early Paleozoic.

Pre-Rift Sediments

The Mesozoic paleogeographic reconstructions within the Red Sea relate mostly to the Tethyan Sea and the passive margin development of Gondwanaland (Fig. 2.4). To the north in Israel and Egypt Jurassic shallow water limestones and marls interfinger with Nubian type sandstone, and mark the Tethyan shoreline, which trends nearly at right angles to the Red Sea trend. The Jurassic sediments are also found in the shelf sequences of central Arabia, and in the south, Jurassic sediments are found in Yemen and along the coastal plain of Saudi Arabia just south of Jizan. In Ethiopia the Jurassic shallow water sediments extend inland just north of Addis Ababa. This Jurassic transgression covered all of the horn of Africa and to the north extended across the present day Red Sea trend at a high angle, so there is no sedimentological evidence that the Red Sea Basin had begun to evolve into its present configuration at this time (Fig. 2.4).

These Jurassic sediments represent shallow water sedimentation, and inter-layered sandstone and shales mark periods of minor regression as these units were deposited. Swartz and Arden (1960) have reported on antipathic relationships between transgression and regression between the northern Red Sea area and the southern Red Sea area. A sequence of shallow-marine limestones were deposited within the countries surrounding the southern Red Sea Basin. These shallow-marine platform carbonate deposits are Late Jurassic, and are represented in Ethiopia by the Antalo formation and in Saudi Arabia and Yemen by equivalent marine limestones called the Amaran formation (Hughes and Beydoun, 1992). Mafic igneous activity is nearly non-existent within these Jurassic sequences surrounding the Red Sea, supporting the idea that the deposits represent shallow seas rather than being the result of extension along a newly developed continental margin.

Early Cretaceous emergence prevailed in the northern Red Sea and transgression continued in Ethiopia, the horn of Africa, and Yemen (Swartz and Arden, 1960; Greenwood and Bleackley, 1967; Grolier and Overstreet, 1978). However, the Cretaceous sediments in the southern Red Sea area in Yemen and Ethiopia consist mainly of sandstone and conglomerates deposited in a continental or shallow marine environment. In nearly all exposures, these sandstones are overlain by the Trap Series, and therefore probably represent the last sediments of the southern Red Sea area unaffected by volcanic and tectonic characteristics of Red Sea Basin evolution (Fig. 2.1).

The clastic Cretaceous sediments signal a regional emergence centered around the southern Red Sea in Yemen, Arabia and Ethiopia. During the Late Cretaceous, the Hail Arch, a broad uplift trending north–northwest developed in central Saudi Arabia (Powers et al., 1966) (Fig. 2.5). The Hail Arch changed the generally east–

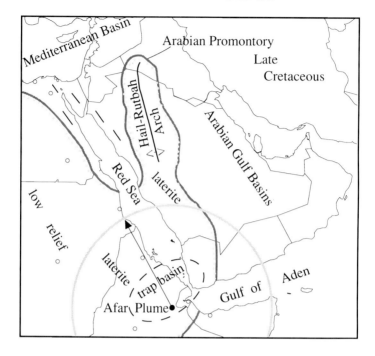

FIG. 2.5. Paleogeographic reconstruction of the Tethyan shoreline for the Late Cretaceous. The dots along the bold line indicate landward areas; the large circle indicates the possible uplift influence for a hot spot centered in the Afar. No palinspastic reconstruction of the plates is given and only the offset along the Dead Sea Rift is shown.

west trending Tethyan Sea shoreline, producing a separate Mediterranean seaway to the west, which was accentuated by north–south graben structures on both flanks (Bender, 1975; Powers et al., 1966). This seaway transgressed southward across Egypt (Said, 1962), forming an embayment as far south as Jiddah (Karpoff, 1955; Brown, 1970; Brown et al., 1989), and extended into the Sudan where Cretaceous marine deposits are present on Maghersum Island (Carella and Scarpa, 1962; Bunter and Abdel Magrid, 1989a,b). The Jiddah embayment had its eastern boundary along the Hail Arch where it extends southward into the Arabian shield, forming shallow estuarine deposits.

The sediments within this embayment consist of fossiliferous mudstones, ferruginous mudstones, dolomites, sandstones, and cherts, with some phosphatic-rich zones (Karpoff, 1955; Carella and Scarpa, 1962; Madden et al., 1980). The sediments suggest shallow water environments typical of lagoons or estuaries, and at the Jabal Umm Himar deposits in Saudi Arabia described by Madden et al. (1980) a tropical climate is indicated by the presence of fossil sharks and catfish. The abundance of oolitic ironstones indicates a lateritic source which had a very low relief. Laterites and paleosols are preserved under the earliest lava flows in Saudi Arabia and Yemen (Overstreet et al., 1977), and laterite has been identified under the earliest basalt flows in Ethiopia (Coleman, unpublished work); this same laterite surface has been reported in the Sudan (Delaney, 1954).

These esturine deposits in Sudan are considered to be Uppermost Cretaceous transitional to Paleocene (Carella and Scarpa, 1962); in the Jiddah area, Karpoff (1955) and Brown (1970) report Maestrichtian to Eocene areas. The Jabal Umm Himar deposits contain a rich fauna of vertebrates and invertebrates, and, equalize according to Madden et al. (1980), indicate a Paleocene age with the restriction that they could be no older than Late Cretaceous and no younger than Middle Eocene.

The Late Cretaceous–Paleocene embayment is the first stratigraphic evidence of an older basin whose axis was parallel to the present Red Sea; however there is no evidence that structures within the basin localized this broad embayment (Fig. 2.5). Regression of the Jiddah embayment during the Early Eocene has been documented in the Sinai, Gulf of Suez, and Egypt (Tromp, 1950; Said, 1962; Madden et al., 1980; Sellwood and Netherwood, 1984; Said, 1990). In the Gulf of Suez, deposition of nummulitic limestones continued uncontrolled by clysmic trends up to the end of the Eocene (Sellwood and Netherwood, 1984; Mitchell et al., 1992). In the northern Red Sea area there are erosional intervals beginning in the Eocene and extending up to the lower Miocene. In the mid-Red Sea area nonfossiliferous continental red beds are found in the Hamamit formation in the exploratory well at Maghersum Island, Sudan and in the Shumaysi formation near Jiddah, where fresh water sediments containing oolitic iron ores in deltaic or lacustrine deposits are present.

These sediments mark an interval of fluvial deposition whose partial source is the widespread lateritic and paleosol surface characteristic of the Arabian–Nubian shield in the Red Sea area. These reworked laterite deposits give rise to the high iron oxide contents (red beds) in the continental fluvial deposits preceding and during the initial stages of rifting. It will be shown later that these continental red beds were probably deposited in graben-like structures developed parallel to the Red Sea axis. Early extension related to the north trending Hail Arch may have initiated the deposition of these continental red beds, but large-scale igneous activity was not apparent until after they were deposited.

Syn-Rift Sediments

No consensus exists concerning the initial stage of rifting in the Red Sea and which observations and data should be used to establish the beginning of rifting. In this book, rifting is considered to be quite distinct from ocean floor spreading. Using the sedimentary history, it is possible to establish sequences that result from the growth of a sedimentary deposition center. It is assumed that some sort of tectonic event is needed to form a sedimentary basin and that at least some volcanic events may be related to the Basin tectonics. Using the recent compilation of the World Bank group (O'Connor, 1992), I will divide the Red Sea Basin sedimentary history into four separate time units which can be recognized throughout

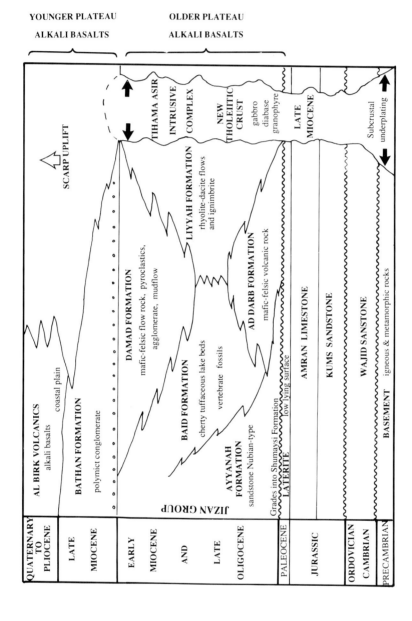

Fig. 2.6. Schematic stratigraphic section showing relationships for the Middle Tertiary Jizan group and Tihama Asir intrusive complex within the exposed rift zone near and around the Jabal Tirf area. Adapted from Hadley et al. (1982;) and Schmidt et al. (1983). Note that the Tihama Asir complex is developing screens of new igneous crust in the Late Miocene.

the Basin (Hughes and Beydoun, 1992): (1) initial rift continental sediments, (2) syn-rift marine sediments, (3) evaporites, (4) post-evaporite marine sediments.

Initial Rift Continental Sediments.

As noted earlier, the final pre-rift event recognized in the Red Sea area was deposition of clastic continental red beds. In the Jiddah area these beds are deposited in a series of grabens parallel to the Red Sea axis. In the Shumaysi formation, which contains oolitic ironstones and continental clastics, pollen in the oolitic facies suggest an Eocene age (Moltzer and Binda, 1981), and Oligocene molluscan shells are present in the upper Shumaysi (Al-Shanti, 1966). Schmidt et al. (1983) propose that the Ayyanah formation consisting of continental sandstones, located near Al Qunfudhah at 19° N, is equivalent in age. In the Sudan, Carella and Scarpa (1962) correlate the Hamamit formation with the Shumaysi formation, a correlation that has been followed by others (Beydoun, 1989) (Fig. 2.1).

On the Saudi Arabian coastal plain a series of disconnected Tertiary outcrops of sediments and volcanics extends from Jiddah to the Yemen border and is included within the Jizan group (Brown et al., 1989, Schmidt et al., 1983). These rocks represent a hetrogenous group of volcanics, pyroclastics, and lacustrine sediments that represent the earliest stage of Red Sea rifting at this latitude (Schmidt et al., 1983, Hadley et al., 1982) (Fig. 2.6). The lowermost member is represented by the Ayyanah formation, which consists of quartz sandstone and quartz pebble conglomerates, derived mostly from the underlying Precambrian basement, resting on a saprolite developed on the surface of the basement (Hadley et al., 1982; Schmidt et al., 1983). These continental Nubian type sandstones are variable in thickness (20 to 40 m), interdigitate with the upper part of the Shumaysi formation near Jiddah, and are overlain by the Baid and Ad Darb formations in the Jizan area (Hadley et al., 1982; Schmidt et al., 1983). The Baid formation consists of lacustrine facies represented by chert, tuffaceous siltstone, claystone, and shale. These deposits are considered to have been derived from silicic pyroclastics deposited in freshwater lakes (Schmidt et al., 1983). These thinly laminated deposits appear to be cyclic and graded, perhaps representing individual volcanic pyroclastic eruptions on the eastern side of the Red Sea (Schmidt et al., 1983). They are present in sparse discontinuous outcrops ranging in thickness from 150 to 300 m, and where exposed, rest on the Ayyanah sandstone or the Ad Darb volcancis or, in some cases, appear to interdigitate with both of these units (Fig. 2.6). This stratigraphic situation suggests a series of disconnected freshwater lake basins developed in the early stages of rifting.

Significantly, the Baid formation contains an important assemblage of fossils, consisting of freshwater pelecypods and gastropods, abundant fish scales and bones, plant debris and vertebrate bone fragments. The discovery of a jaw fragment of a hippo-like even-toed ungulate (*Masritherium, Artiodactyla, Anthracotheriidae*) related to a North and East African genus of Early Miocene age (Madden et al., 1983) indicates a setting analogous to present-day East African rift valleys. Contemporaneous interlayered subaerial volcanics are divided into the upper unit (Damad formation), middle unit (Liyyan formation), and lower unit (Ad Darb formation), representing large eruptions of rhyolitic to andesitic material (Schmidt et al., 1983). These bimodal volcanic rocks consist predominantly of green tuff, agglomerates, welded tuff, and lava flows. All of the units have undergone hydrothermal alteration, indicating the presence of abundant fluids provided by freshwater lakes into which these eruptions extruded, filling the newly-formed pull-apart or graben-like basins.

The volcanic units vary in thickness from 500 to 1500 m and their areal distribution is related to the individual eruptive centers. Intrusive sills and cross-cutting dikes from the Tihama Asir complex have been dated between 19 and 20 Ma and place an upper limit on the age of the Jizan complex (Schmidt et al., 1983; Brown et al., 1989). These important new discoveries on the eastern margin of the southern Red Sea reveal for the first time that this part of the Red Sea did not consist of a large Nubian–Arabian dome but was instead a continental rift valley or a series of graben-like depressions receiving continental clastics and bimodal volcanic flows and pyroclastics. The presence of freshwater invertebrate shells in these sediments indicates lacustrine deposits, probably similar to present-day East African lakes.

As mentioned earlier, inital rift outcrops of the Hamamit formation are present on the Sudan coastal section and have also been reported in the Maghersum No. 1 exploration well, where they rest unconformably on the Mukawar formation, representing the southernmost transgression of Tethyan ocean (Carella and Scarpa, 1962; Bunter and Abdel Magrid, 1989a,b). The Hamamit formation is 220 to 260 m thick and is overlain by the Maghersum formation of the Middle to Lower Miocene (Carella and Scarpa, 1962; Bunter and Abdel Magrid, 1989a) (Fig. 2.1).

The lithology of the Hamamit formation is similar to that of the Shumaysi formation near Jiddah and the Lower Miocene–Oligocene red beds of the Suez area near Wadi Tayiba section (Carella and Scarpa, 1962; Sestini, 1965; Sellwood and Netherwood, 1984). These deposits are non-marine fluvial deposits consisting of red-brown to green, coarse to fine grained sandstones cemented with iron oxides and containing lenses of conglomerates with well-rounded clasts of Precambrian basement rocks (Carella and Scarpa, 1962; Sestini, 1965). A single basaltic flow (~30 m) is interlayered with the sandstones, signaling the beginning of rifting in the Sudan coastal area.

Along the modern Red Sea coastline in Ethiopia from Anfile Bay to Massawa, the Dogali formation consists of a sequence of nonfossiliferous sandstone and conglomerate interbedded with basaltic lava, breccia, and tuff that reaches a thickness of nearly 3000 m (Filjak et al., 1959). The Dogali sediments along the Ethiopian shore rest unconformably on Mesozoic marine sediments and represent the initiation of rifting (Savoyat and Balcha, 1989). The Dogali formation has also been encountered in several exploration wells offshore (Lowell and Genik, 1972). The Dogali series is considered to be of the Oligocene period (Savoyat and Balcha, 1989), and can be correlated with the Jizan group in Saudi Arabia and the Hamamit formation in Sudan (Fig. 2.1).

The lithology is quite variable, consisting primarily of fluvial continental deposits interlayered with lacustrine sediments containing fish scales and bones, similar to the Baid formation near Jizan. Interlayered pillow lavas and tuffs indicate rift activity contemporaneous with the earlier volcanic eruptions recorded in the Afar (~24 Ma.) (Tierclelin et al., 1980). These sedimentary units (up to 2500 m) encountered in some of the exploration wells offshore grade upward into the marine and deltaic sediments of the Habab formation. The stratigraphic relations indicate an age between the Eocene and Early Miocene periods for the initial rift sediments and volcanics (Savoyat and Balcha, 1989).

In the southern Red Sea the emergent laterized Nubian–Arabian shield surface was partially covered by the Trap Series volcanics during the Early to Middle Oligocene period. These flows interlayered with ignimbrites in Yemen and eventually attained a thickness of 300 m along the present-day Red Sea rift margins by the Early Miocene period (~19 Ma). The basal flows of the Yemen Traps rest conformably on the upper Medj-zir member of the Tawilah formation, and K/Ar radiometric ages of these lavas are ~26 Ma. The Medj-zir sediments are continental fluvial fine grained sediments whose upper surfaces have been affected by deep weathering, with the development of paleosols very similar to those present under the Sirat volcanics near the Yemen–Saudi Arabia border (Coleman et al., 1983; Menzies et al., 1990).

Initial rift sediments are present along the northwest shore of the Red Sea from Ras Banas to the Gulf of Suez and consist of a thick sequence (200 to 500 m) of continental clastics containing conglomerate units derived from the underlying Tethyan marine sediments (Thebes formation) and Precambrian basement (Purser

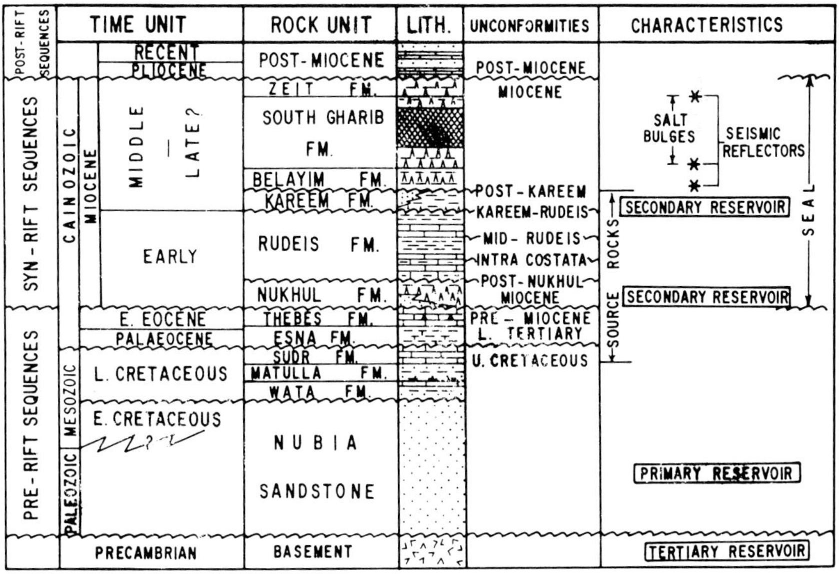

FIG. 2.7. Generalized stratigraphic section for the southern Gulf of Suez, taken from Helmy (1990).

and Hotzl, 1988; Purser et al., 1990b). Low angle tilting of the Thebes formation caused by normal faulting was followed by continental alluvial deposition and extrusion of interlayered basalts (~24 Ma). These deposits consist of red-brown mudstones and sandstones with conglomeratic units deposited in alluvial fans and channels, accompanied by fine grained flood plain sediments (Purser and Hotzl, 1988; Purser et al., 1990b). These continental sediments have yielded no fossils but the stratigraphic position and K/Ar age of the interlayered basalt suggest an Oligocene to Lower Miocene age. On the northwest edge of the Red Sea, continental sedimentary debris has also been concentrated within strike slip basins oblique to the Red Sea axis (Orszag-Sperber and Plaziat, 1990; Plaziat et al., 1990).

In the Gulf of Suez, a regional erosional surface developed on the Eocene chalks is marked by thin beds of red conglomerate associated with silicic tuff (Sellwood and Netherwood, 1984). Here, there is a rejuvenation of the Nubian–Arabian laterite surface along with normal faulting but confined to the Red Sea–Gulf of Suez basin. Interpretation of the Eocene to Oligocene events in the Gulf of Suez requires a period of uplift and major faulting, with the formation of continental valleys collecting continental clastics contemporaneous with a variety of interlayered bimodal volcanic rocks (Fig. 2.7). On the east side of the Gulf of Suez in the Wadi Tayiba section the Eocene Thebes limestone was eroded and covered by red beds consisting of conglomerate containing limestone and chert pebbles derived from the Thebes formation (Sellwood and Netherwood, 1984). At this location, below the conglomerate, a thin bed of water laid crystal-rich tuff marks the initiation of volcanic activity. This is overlain by a 15 m section of alkali-trachy basalt (~24 Ma) which is similar to the 20 to 25 Ma ages obtained on dikes and flows within the Sinai (Bartov, 1980). The uppermost section of the basalt has pillow structures, indicating submarine eruption (Sellwood and Netherwood, 1984). The overlying Lower Miocene sediments are conglomerates, which fill in a rough topography above the basalts. The matrix of the conglomerates contain Early Miocene benthic foraminifera and *Thalassinoidess* burrows. These relations suggest that an early littoral environment followed the initial volcanic eruptions (Sellwood and Netherwood, 1984). Interpretation of the Post-Eocene events in the Gulf of Suez requires a period of uplift and major faulting, with the formation of valleys collecting continental clastics contemporaneous with a variety of interlayered bimodal volcanic rocks (Fig. 2.7). The presence of the continental to lacustrine alluvial sediments associated with volcanic eruptions along the Red Sea margins indicates that the initial Red Sea Basin was not part of a large Nubian–Arabian dome, but consisted of a continental rift valley or a series of graben depressions receiving continental clastics and bimodal volcanic flows with deposition within fault bounded lake basins.

Syn-Rift Marine Sediments

During this interval, a huge extrusive pile nearly 2000 m thick formed from the Yemen bimodal volcanics. Alkali olivine basalts extruded some 150 km east of the Red Sea continental rift, forming extensive plateau basalts north from Yemen to Jordan and Syria (Coleman et al., 1983). At this stage, block faulting and extension defined the shape of the Red Sea and Gulf of Suez. At Jizan, the Tihama Asir complex dike swarms and gabbros cut the earlier-formed Jizan group and developed screens of new crust between extending and faulted continental crust (Fig. 2.6). In the Early Miocene stage, lowering of the global sea level isolated both the Red Sea rift as well as the Mediterranean, and the Red Sea received only incremental surges of sea water from its Mediterranean connection.

The initial rift and syn-rift sedimentary history is difficult to correlate because many exploratory wells did not penetrate this deep, and the continental derived submarine cones represent locally derived alluvial sources. The syn-rift volcanic extrusions within the basin are disparate, and represent only a small volume within the basin. Syn-rift marine sequences older than 20 to 25 Ma are present in the southern Red Sea and are characterized by the Habab formation in the Ethiopian section, the Maghersum formation in the Sudan section, and the "Infra evaporite series" in the Saudi Arabian section. All of these are time equivalents of the Rudies formation within the Gulf of Suez (Fig. 2.1). These syn-rift marine sequences of the Red Sea and the Gulf of Suez are mainly fine grained clastic sediments deposited in a low energy environment in the central part of the basin, but coarsen near the shoreline source areas where high energy submarine fans develop (Hughes and Beydoun, 1992). The central part of the Red Sea is generally barren of microfauna, indicating a suboxic to anoxic bottom condition (Crossley et al., 1992). By Middle Miocene evidence from the wells indicate that moderately deep marine conditions extended throughout the Red Sea with deep marine conditions present only in the southern Red Sea (Crossley et al., 1992). Early Miocene detrital deposits form important local buildups, which form lobate submarine fans with steep dips (Purser et al., 1990b). These fans coalesce seaward and form the Middle Miocene syn-rift marine clastic wedge. These marine sediments were deposited on and around the structural highs developed by normal faulting during extension. Along the northwest shoreline of the Red Sea numerous reefs developed on the earlier deposited continental initial rift, continental red beds or on the marine depositional cones (Purser et al., 1990b). Evidence of the salinity crisis to come is found within the Karem formation in the Sudan and Egyptian sectors, where the Markha member indicates restricted conditions with hypersaline deep marine deposits (Hughes and Beydoun, 1992) (Fig. 2.1).

Evaporites

These incursions of seawater from the north and perhaps from the south produced a Miocene evaporite series up to 3 to 4 km thick in the Red Sea (Heybroek, 1965). The evaporite sequences have been correlated throughout the Red Sea, and the early stages are typical of the Belaymin formation on the Egyptian side, which shows periodic hypersaline deposition interlayered with deep marine clastic sediments (Hughes and Beydoun, 1992). The main evaporite series in the Gulf of Suez is represented by the South Gharib and Zeit formations, which are considered equivalent to the Dungunab formation in the Sudan "Evaporite Series" in Saudi

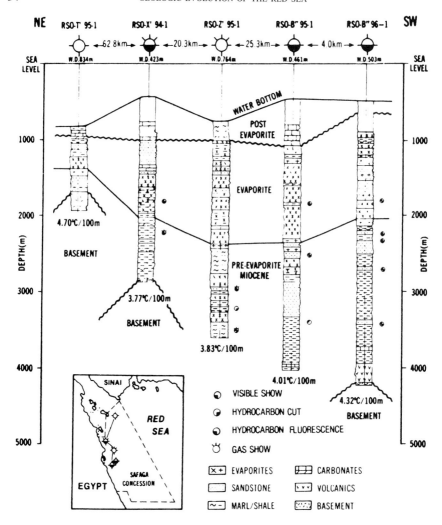

FIG. 2.8. Generalized stratigraphic correlation of exploratory wells in the Safaga concession in the northern Red Sea (Miller and Barakat, 1988).

Arabia, the Amber formation in Ethiopia, and the Salt formation in Yemen (Figs. 2.1 and 2.7). The salt water incursions were cyclic (~10 m in thickness), starting with deposition of clays and carbonates containing pelagic fauna of globigerinids and pteropods with Mediterranean affinities. Evaporation of the basin produced basinal evaporites and prograding sabkhas along the margins, followed by a new phase of flooding from the Mediterranean (Sellwood and Netherwood, 1984). Continued volcanic activity produced interlayered volcanics and clastics from fault scarps and prograded across the evaporite cycles, producing a mixed facies of continental clastics and sabkhas. During this period only modest uplift had taken place along the margins of the Red Sea, as shown by the balancing of evaporite deposition against clastic sequences, which are less than 3 km thick.

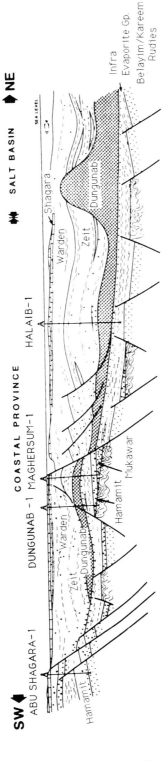

FIG. 2.9. Schematic structural cross-section of the north central Red Sea extending from the western Sudan margin, showing the approximate positions of exploratory oil wells. Taken from Mitchell et al. (1992), with permission.

35

Two models have been used to explain the evaporite deposits of the Red Sea: (1) shallow water evaporation with associated sabkhas; (2) the deep basin model which maintains near-constant sea level with surface precipitation of "rafts" or clumps of evaporite (Hughes and Beydoun, 1992).

Evaporites penetrated by deep-sea drilling project, Leg 23B (Stoffers and Kuhn, 1974) have alternations of halite and anhydrite, and dolomitic black shales suggesting cyclic concentration of brine, which is very characteristic of shallow-water environments. Such observations from Leg 23B of the deep-sea drilling project (DSDP) may not hold for the whole Basin as there is abundant evidence that the Red Sea trough contains deep water marine sediments, as demonstrated by the pelagic assemblages present. Stromatolites interbedded with halites from the Egyptian coastal sequence and the presence of potassic salts detected by well logging indicate shallow-water salt evaporation (Crossley et al., 1992). All of these observations can be reconciled if it is assumed that episodic batch-filling of the basin is followed by evaporite draw-down within a deep basin (Crossley et al., 1992).

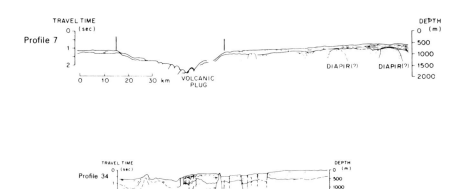

FIG. 2.10. Selected reflection seismic profiles from the Red Sea. The S-reflector is indicated as a heavy dark line and the vertical ticks indicate boundaries of the axial trough. Travel times are two-way (Ross and Schlee, 1977).

Exploratory wells drilled in the Red Sea have reached the evaporite deposits and in several instances have penetrated completely through (Figs. 2.8 and 2.9). In some cases, drilling reached granite or basalt without ever encountering evaporites (Stoffers and Ross, 1977; Beydoun, 1989). The evaporites in the Gulf of Suez and the Red Sea form a sequence nearly identical in age with the Mediterranean evaporite deposits (Hsu and Bernoulli, 1978; Cita and Wright, 1979). Brown (1970) shows that the evaporite basin of the Red Sea nearly conforms to the present Red Sea shoreline; however, correlations between close spaced wells that have penetrated the salt indicate a tremendous variation in thickness. Much of the variation may result in post-depositional diapirism (Mart and Ross, 1987) and flow of salt weakened by the high heat flows present within the axial part of the Red Sea (Girdler and Styles, 1974) (Fig. 2.9).

Interlayered shales contain globigerina, whose age indicates that the evaporites encompassed most of the Miocene, accompanied by continental stretching and subcrustal intrusions of mafic magma as the basin enlarged. Apparent thinning of the evaporates towards the axial trough indicates this slow extension and enlargement. However, there are no deep wells penetrating the salt along the edge of the axial trough that can verify the nature of the crust during the Late Miocene. In the Late Miocene, the Gulf of Aden began to open, forming oceanic crust, and at the same time movement on the Dead Sea accommodated this movement along a major strike slip fault (Girdler and Styles, 1974).

Post-Evaporite Marine Sediments

At the beginning of the Pliocene (~5 Ma) evaporite deposition ceased and is marked by a major unconformity in the Red Sea and Gulf of Suez (Ross and Schlee, 1973; Sellwood and Netherwood, 1984) (see Fig. 2.1). Overlying the faulted and deformed evaporite is a sequence of marine sediments that are characteristically nanno oozes, including silty clays, chalk and dolomites, all of which contain pelagic fossils of Indian Ocean affinities (Stoffers and Kuhn, 1974; Stoffers and Ross, 1977). Seismic-reflection profiles show a widespread truncation of deformed reflectors in the evaporite section by the overlying marine sediments (Fig. 2.10). This discontinuity is now interpreted as a major unconformity, marking the initiation of sea floor spreading in the axial trough of the Red Sea (Young and Ross, 1970; Ross and Schlee, 1973; Coleman, 1984a). This unconformity is present throughout the main trough but does not extend into the axial trough.

Subsidence from the very late Miocene to the present time produced two important depocenters in the Ethiopian sector: the offshore salt basins and the coastal basin. The Desset formation rests on the evaporite sequence and at its base (300–2000') consists of deltaic to sabkha-continental environment. The middle portion of the Desset formation consists of sand–shale sequence infilling the lows between the diapiric salt domes. Overlying these infillings is a deltaic sand–shale sequence of shallow-marine shelf facies rich in carbonate lenses near the salt domes. Unconformably overlying the Desset formation and the salt diapirs is the

Dhunishub formation (200–2000') with reefal limestone (Savoyat and Balcha, 1989). Salt diapirs formed during the Late Pliocene with gliding of the domes to the northeast have formed a salt wall within the west boundary of the basin (Savoyat and Balcha, 1989). The Dhunishub seals all of the salt crests in the basin except in the Dahlak Islands, where salt is still moving upwards, similar to the situation to the east in the Farasan Islands.

In other parts of the Red Sea, near-shore units equivalent to the Dhunishub formation are characterized by sand-dominated alluvial fans and coastal plain sediments, with deltas developing on the western side of the Red Sea near major inherited fault structures at the Tokar Delta and a large delta north of Port Sudan. Carbonate buildups related to reef development are more abundant on the eastern shoreline of the Red Sea. The central part is characterized mainly by open marine muds (Crossley et al., 1992).

The lack of Miocene salt deposits and Pliocene marine sediments in the axial trough has been established by drilling on Leg 23 of the DSDP (Ross and Schlee, 1973). In the axial trough, a thin veneer of sediments (probably not more than 28,000 BP) is often interlayered with metalliferous deposits in the deeps associated with hot brines (Backer and Schoell, 1972). The sediments in the axial trough are Pliocene or younger, but Post-Miocene salt flow invading the axial trough has often disrupted the original stratigraphy (Girdler and Whitmarsh, 1974). Pliocene–Pleistocene rates of deposition in the southern Red Sea Basin were distinctly higher than those observed in the northern parts of the basin (Whitmarsh et al., 1974). On the Saudi Arabian coast, between latitudes 21° and 13° N, the Bathan formation, consisting of around 200 m of polymictic conglomerates (with clasts up to 2.5 m), is deposited above the Jizan group and the Tihama Asir igneous complex. Basinward, a correlative unit (unnamed) from the Mansiyah-1 well consists of 2000 m of coarse clastic debris overlying the evaporites. On the Sudan coast Montenant et al.(1990), Sestini (1965), and Carella and Scarpa (1962) describe Pliocene clastic units of coarse sands and gravels resting on Miocene evaporites. Coarse clastics with gypsum interbeds are found above evaporites in exploratory wells along the Ethiopian coast (Hutchinson and Engels, 1970). Within the Afar Depression the Danakil Red Series represents an erosional event related to the uplift of the Ethiopian and Somalian plateau and has been dated as Miocene–Pliocene (CNR/CNS Afar Team, 1973; Tierclelin et al., 1980).

There is a remarkable coincidence of these Pliocene clastic units with the major Pliocene–Miocene unconformity within the margin of the Red Sea Basin. The magnitude of this sudden shift from a closed evaporite basin, with only moderate clastic input from the surrounding hinterlands, to open ocean sediments interlayered with coarse clastics from rising highlands, records a profound change in sedimentation and tectonic setting. This event has also been recorded in the Mediterranean basin, where a distinct unconformity is present throughout the basin (Cita and Wright, 1979). In the Gulf of Suez, this Miocene–Pliocene unconformity marks a tectonic event that tilts Pliocene gravels up to 5°. The stratigraphic record for the Miocene–Pliocene Red Sea event signals the beginning of ocean sea-floor spreading in the axial trough and rapid uplift of the Red Sea margins in the southern part of the basin (Evans, 1988). Shallow drilling for water on the Arabian shoreline encounters Holocene coral reefs and interbedded sands. The Holocene coral reefs are uplifted and deformed in the north, whereas continued high sedimentation rates in the south continue to cover these reefs (Coleman et al., 1983).

Sedimentary Constraints

The sedimentary history of the Red Sea evolution provides a number of basin-wide episodes that constrain kinematic models of plate motion:

1. Evidence for Paleozoic or Mesozoic sedimentation trends that follow the proto-Red Sea structure is lacking.

2. Intermontane continental red beds and fluvial deposits follow Red Sea graben-horst trends in the Oligocene period. Presence of these deposits negates the often invoked Nubian–Arabian dome prior to rifting.

3. Extension and widespread volcanism in the Early Miocene period defines the structural shape of the Red Sea depositional basin.

4. Evaporite deposits are coeval with the Mediterranean salinity crisis and confirm the closed-basin nature of early extension in the Red Sea.

5. A major tectonic event is recorded by the Pliocene–Miocene unconformity that marks the change from evaporite deposits to open ocean deposits containing pelagics from the Indian Ocean.

6. Coarse continental derived clastics invaded the Red Sea Basin in the Pliocene period and mark nearly synchronous flank uplifts for the Arabian, Yemen, Somalia, and Ethiopian Plateaux.

7. Sedimentation rates in the Red Sea Basin follow present day configuration of the regional uplifts, high in the south and low in the north.

3

Volcanic History

It is now generally considered that the development of the Red Sea was related to a diversity of volcanic activity, and that these eruptive products were not all synchronous. It is important to study these rocks to establish their eruptive history with the evolution of the Red Sea Basin. The volcanic activity can be geographically divided into more or less coherent regions: (1) the northern termination of the East African rift system, consisting of the Afar Depression transitional volcanics and the surrounding Ethiopian Plateau basalts, where it abuts against the southwestern margin of the Red Sea; (2) the Yemen volcanic plateau on the opposite side of the Red Sea, which is an asymmetric counterpart to the Afar Depression, consisting of a thick pile of bimodal volcanics; (3) the western Saudi Arabia Plateau basalts, forming separate and distinct volcanic centers; and (4) the axial trough of the Red Sea, where active sea floor spreading is forming new oceanic crust in the southern and median sections. All of these areas are now sufficiently well known to enable the establishment of their eruptive histories. Details about historical eruptions within the Red Sea area can be found in Simkin et al. (1981).

The Afar Depression

The Afar Depression is roughly triangular in shape and is bordered on the west by a scarp consisting of Precambrian Pan-African basement overlain by the very thick sequence of Ethiopian Plateau lavas (Mohr, 1970; 1983; 1989). On the east, the Danakil block, consisting of basement rocks with a cover of Mesozoic marine sediments, is only sporadically covered by Tertiary volcanics. The volcanic and sedimentary sequences are flat lying within the depression, and the older stratigraphy is not exposed at the surface. The controversy regarding the Afar results from the lack of older exposed stratigraphy, and it is not certain whether the marginal units extend into and make up part of the basin or whether new oceanic-like crust is present underneath the Depression (Barberi et al., 1970; Barberi and Varet, 1975; Mohr, 1978). The oldest sediments to be deposited in the Afar Depression, the "Red Series," consist mainly of terrestrial alluvial deposits that interdigitate with lacustrine and marine clays. K/Ar ages on basalts interlayered with sediments give ages ranging from 24 to 5.4 Ma, and the sediments reflect rapid changes in tectonism and volcanism as the basin formed (Tierclelin et al., 1980).

The most extensive eruptive rocks of the Afar Depression are the stratoid basalts (Trap Series) erupted from marginal fissures that parallel the marginal

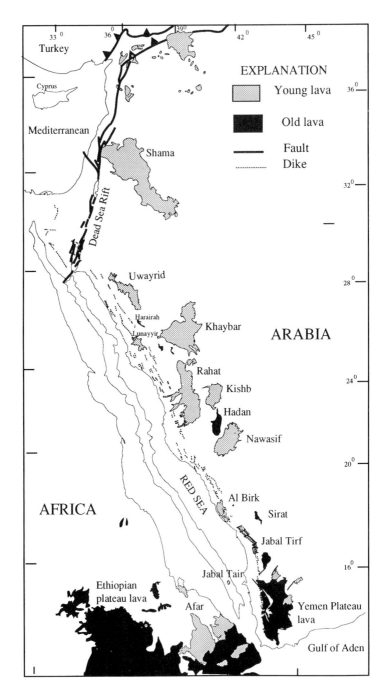

Fɪɢ. 3.1. Index map of Tertiary volcanic fields associated with structures of the Red Sea. The old lavas are generally between 5 and 25 Ma, and the young lava is less than 5 Ma. The dikes are ~21 to 25 Ma and offset by the Dead Sea Rift, which fixes its age to less than 21 Ma.

fault systems (Mohr, 1978). Eruption of the stratoid series began in the Late Oligocene period and continued sporadically up to the Late Miocene. These basalts are transitional types, consisting mainly of hawaiite-icelandite often interlayered with trachytes and pantellerites. These lavas have an evolved low-pressure fractionation trend along with iron enrichment and it is unlikely that any of these rocks represent magmas characteristic of mid-ocean ridge basalts (MORB) (Barberi et al., 1972; Mohr, 1978).

The most spectacular manifestation of volcanism is localized along the axial ranges that have a northwest to southeast trend and whose eruptive history is confined to the Quaternary (Barberi et al., 1972; Tazieff et al., 1972). These eruptions developed an apron up to 40 km wide and extending more than 100 km along the axis (Mohr, 1978), with more recent extension producing graben structures and eruption of basalts along fissures. Although these rocks could be considered similar to MORB, their high potash contents and evolved nature set them apart quite distinctly from the basalts in the axial trough of the Red Sea. Eruption of central volcanoes in the axis show advanced fractionation, with the formation of abundant silicic trachytes and pantellerites (Barberi et al., 1974b). The early development of fissure eruptions evolving to shield volcanoes and finally to stratovolcanoes suggests that the slow extension rate (much lower than most oceanic spreading rates) allowed extensive fractionation and the production of transitional eruptive products rather than the subalkaline tholeiites characteristic of mid-ocean spreading centers.

Yemen Volcanics

The Tertiary volcanic rocks of the Yemen have often been referred to as the Trap Series (Lipparini, 1954; Shukri and Basta, 1954; Geukens, 1966; Mohr, 1983; Manetti et al., 1991). These rocks form a high plateau area whose eroded western margin forms a scarp facing the Red Sea. This volcanic pile averages more than 1200 m in thickness (Fig. 3.1), and consists of alternating flows of basalt inter-bedded with acid effusive ignimbrites that range in composition from rhyolite to comendite. It is estimated that the ratio of basalt to acid volcanics is approximately 1:1 (Capaldi et al., 1983).

The basal flows of the Trap Series rest on shallow marine Medj-zir sandstones and conglomerates considered to be of the Paleocene system (Geukens, 1966). The basin containing these Paleocene sediments appears to be thickening westward towards the present axis of the Red Sea. The abundance of ferruginous weathering products within the upper beds indicates that extensive paleosol developed prior to the eruption of the lavas, providing evidence that pre-eruptive doming was absent in the Yemen sector (Geukens, 1966). The pre-eruptive surface appears to have been quite irregular, with some sections only a few meters thick along the eastern borders but thickening up to 2000 m within the central portion of the shallow marine basin (Geukens, 1966). The Red Series in the younger graben basin of the Afar triangle is considered to be Lower Miocene and unrelated to the Medj-zir basin sediments (Tierclelin et al., 1980).

There is sparse knowledge concerning the eruptive history of the Yemen Trap Rocks, but in general the acid effusive rocks are rhyolite tuff or ignimbrites interlayered with obsidians, exhibiting variations in color from pale reds to greens. These silicic units are reported to develop uniform thicknesses up to 10 m and may reach an aggregate thickness of 300 m. The interbedded mafic volcanics are mainly alkaline basalts and hawaiites with some mugearites and trachytes. These mafic flows may reach thicknesses of up to 100 m and exhibit characteristic columnar to concentric cooling joints (Capaldi et al., 1983). A hiatus in eruptive history within the Traps is manifested by lacustrine interbeds, continental sandstones and paleosols, all indicating continued basin-like topography during the eruptive history (Geukens, 1966).

Radiometric K/Ar ages on the Trap volcanics reveals a range in age between 20 and 30 Ma (Civetta et al., 1978). Later studies by Capaldi et al. (1983) revealed that some of the basalts in northern Yemen are approximately 10 Ma and represent alkaline and hawaiite lavas. These intermediate ages reveal that there have been minor eruptions between the older Trap Series and the Quaternary and Recent alkali basalt volcanism present in the Sana and Marib areas. Additional recent volcanics are present in the Dhamir and Rida areas, where fissure eruptions evolve into central volcanoes and rhyolite domes are associated with explosive craters (Civetta et al., 1979). Comparisons of the recent Yemen alkaline basalts with those from the Eastern Afar margin, the Red Sea islands and southern Yemen indicate a similar petrographic nature with geochemical signatures that suggest deep mantle sources with a low percentage of mantle melting (Civetta et al., 1979). Further comparison with the Al Birk volcanics in Saudi Arabia supports the idea of modern off-axis, limited deep mantle partial melting.

Plateau Basalts (Arabia)

One of the largest areas of predominantly alkali olivine basalts is found on the Arabian side of the Red Sea and situated within Saudi Arabia, Yemen, Jordan, and Syria (Fig. 3.1). More than twenty separate volcanic fields are known and the estimated volume of eruptive material is 10^3 to 10^5 km^3. The older volcanic fields trend approximately parallel to the present Red Sea axis and the younger harrats have eruptive centers that trend between N 20°W to 19°E. In the following discussion of these lavas, the older Oligocene–Miocene centers will be considered first, followed by the Pliocene–Holocene eruptions, as there are important differences between these two periods.

Older Harrats

Harrat As Sirat (750 km^2) is situated near the edge of the Red Sea erosional scarp along the Yemen border and may be the most northerly exposure of the much more extensive Yemen Traps (Capaldi et al., 1983; Du Bray et al., 1991). These lavas were extruded on extensive paleosols (lateritic) that were derived from the underlying Precambrian basement rocks (Overstreet et al., 1977) (Fig. 3.2). These

FIG. 3.2. Edge of scarp just south of Ursan at ~2500 m. The dark circular area is an erosional outlier of the older Harrat As Sirat basalt flows, resting on laterite (paleosol) developed on Late Precambrian granodiorite. View from helicopter at ~120 m looking north (1971).

laterites are correlated with a similar laterite that underlies the basal flows of the Ethiopian Plateau (Mohr, 1971) and a paleosol formed at the top of the Eocene Umm Himar formation northeast of Taif (Madden et al., 1980; Brown et al., 1989). To the south at the Yemen border the basal flows rest unconformably on the Early Paleozoic Wajid sandstone. Numerous volcanic necks exposed by erosion northwest of the Harrat As Sirat have a trend that is parallel to the axial part of the Red Sea. These volcanic necks are up to 15 m in diameter and exhibit columnar cooling joints normal to their contacts with the basement rocks. These feeder necks extend nearly 65 km to the northwest and outline a much more extensive eruptive area that may have been present prior to erosion.

The Harrat As Sirat consists of 35 separate flows with a maximum total thickness of 580 m (Coleman et al., 1983; Du Bray et al., 1991). Individual flows are 3 to 20 m thick and characteristically have columnar joints in the massive lower and middle sections. Trachyte domes are developed at several points within the harrat but their volume is quite minor. The lack of paleosols or erosion between flows indicates that the Sirat basalts represent an intact eruptive sequence that is 22 to 30 Ma (Coleman et al., 1977; Du Bray et al., 1991). No folding or warping of the Harrat As Sirat flows is apparent but north trending normal faults down to the center of the field indicate a graben-like structure (Du Bray et al., 1991). Flow directions could not be established but the absence of Sirat lavas

FIG. 3.3. Basal contact of the older Harrat Hadan lavas on the east side resting on the Paleocene Umm Himar marine limestone, forming the small bluff in the foreground (Madden et al., 1980). K/Ar ages on basal flows are ~30 Ma. View looking north from helicopter at 175 m, approximately 150 km east of Taif (1981).

overlying Wajid sandstone in the Jabal Tirf area 60 km to the west indicate that a gradient towards the Red Sea probably did not exist during the eruptions.

Harrat Hadan (3700 km²) is an elongated, highly dissected, and eroded area of basaltic flows that forms the eastern barrier of the great Sahl Rakbah (a large open plain). An erosional escarpment in the Taif area 90 km marks to the west the edge of the Red Sea scarp. Harrat Hadan forms a prominent lava plateau, rising some 300 m above the Sahl Rakbah along its west facing escarpment. East flowing wadis have deeply dissected the lava plateau whose eastern flank is subdued as a result of eastward tilting.

The lavas at Harrat Hadan were partly deposited on a paleosol (lateritic), and some of the lower flows are interbedded with mudstone and limestone (Madden et al., 1980) (Fig. 3.3). The paleosol is developed from the Paleocene Umm Himar formation, a marine esturine deposit that formed in a shallow seaway that extended southward from Jordan. The deep erosion of the west side of the Harrat Hadan has exposed numerous north trending feeder necks that form the central axis of the volcanic eruptions (Fig. 3.4). A north trending dike swarm at the northern tip of Harrat Hadan may be related to similar trending faults in the basement to the south, but it is not clear that these north trending structures are related to eruption of the lavas.

The lava flows are typically 8 to 10 m thick with the lower flows most intensely weathered, forming knobby and recessive slopes. The basalts are olivine

FIG. 3.4. Circular feeder neck for the older Harrat Hadan, displaying vertical cooling joints. West side of Harrat Hadan, looking southwest (1981).

transitional and alkali olivine, with some nepheline. Hawaiites are sparse and no interbedded silicic lavas are present. The lower flows have a regional dip of about 10 degrees to the east, whereas the upper flows appear to be horizontal. This unconformity can be seen within the erosional re-entrants along the west side of the harrat. At least some of the uplift and tilting east of the Red Sea scarp took place in the Miocene, as the K/Ar age for the older tilted flows is 27 Ma, and that of the younger horizontal flows is 16 Ma (Arno et al., 1980; Brown et al., 1989).

Harrats Ishara-Khirsat and Harairah occupy a highland area about 90 km inboard from the Red Sea shore and extend along Wadi al Jizl and Hand from 26°15' to 24°30' (Brown et al., 1989). Eroded volcanic necks define a north 30°W trend and the thickness of the flows varies from 30 m to a maximum of 400 m, indicating a moderate topographic relief at the beginning of eruption. Harrat Harairah is thickest at latitude 26°10' N and thins southeastward, whereas Harrat Ishara-Khirsat is thickest at 24° 40' N.

The basalt flows for the separate harrats are similar, and appear to have nearly identical sequences, with olivine transitional basalts at the base grading upward into alkali olivine flows containing small mantle xenoliths. The upper flows are alkali olivine basalts with diktytaxitic textures that may have evolved to hawaiites.

At Harrat Harairah lag deposits of silicified coquina associated with pebbles and cobbles of chert underlie the basal lavas (Fig. 3.5). These deposits are similar

FIG. 3.5. Erosional outliers of Harrat Harairah basalt flows resting on a thin layer (~2 m) of lag deposits consisting of silicified coquina and chert cobbles derived from Paleocene–Eocene marine deposits. Deeply eroded material below the gravels and basalt is Precambrian basement. View looking south approximately 50 km south of Al Ula (1981).

to lag deposits found further north under Harrat Uwayrid, widespread deposits of lag chert in the region of Turayf near the Jordan border, and to those found under Harrat Hadan, and are thought to represent a post-Eocene erosional surface developed upon the shallow-water marine sediments whose age may span from Paleocene to Eocene. At Jabal Antar, in Harrat Ishara-Khirsat, fanglomerate sequences underlying the basal lava flows (Kemp, 1982) have bentonite layers at their graded tops that often contain petrified wood. Preliminary K/Ar ages indicate a range in age from 21 to 28 Ma (Kemp, 1982).

Summary of Older Harrats

The stratigraphic evidence for the older harrats supports the idea that from Paleocene to Middle Oligocene the western portion of the Arabian plate was low-lying and near sea level. There is no evidence that a regional uplift had preceded any of the early eruptive activity. Some tilting can be seen at Harrat Hadan, where an angular unconformity separates the older flows from the younger flows, so in the later stages (Middle Miocene) some uplift is apparent. The older harrats characteristically erupted from central vents fed by pipes (>200 m in diameter) showing an irregular pattern. Dikes were not revealed in areas of deeper erosion.

The lack of fractionation is characteristic as the bulk of the flows are either olivine transitional basalts or alkali olivine basalts. It is rare to see any indication of differentiation of these older lavas, and the magmas are clearly unrelated to the tholeiitic rift magmas.

Younger Harrats

The younger harrats extend from the most southerly part of the Red Sea coast on the Arabian side, from Jizan northward to Amman, Jordan. Nearly all of these harrats have an early history of eruption some time after 15 Ma, extending up to the present time. They form a series of unconnected volcanic centers that occupy the eastern edge of the Arabian plate.

Harrat Al Birk is situated on the Red Sea coastal plain, mainly between Wadi Hali and Wadi Nahb, covering the coastal plain from the foothills to the shoreline (Arno et al., 1980; Brown et al., 1989). The flows form an irregular elongate field of approximately 1800 km². Inland from the main field, two separate cones (Jabal al Haylah and Baqarah) and associated flows may belong to the same volcanic system. Many small cinder cones and associated flows are present southward near Jizan (Fig. 3.6). The lavas cover paleotopographic features such as wadi

Fig. 3.6. Small basaltic cinder cone related to the young eruptive sequence of Harrat Al Birk. Jabal Akwaw situated on the Tihama coastal plain 35 km N of Jizan, view looking westward towards the Red Sea (1971).

valleys and pediment surfaces, and aggregate into a nearly level plateau surface just a few meters above present sea level. At Jabal al Haylah, the lava covers gravel terraces that contain Acheulian implements (W.C. Overstreet, personal communication, 1977). In Europe, Acheulian implements are suggestive of Lower-Paleolithic age between the second and third interglacial periods ~500,000 years (Brown et al., 1989). On the eastern side of Harrat Al Birk, lavas overlie Precambrian basement that is invaded by Tertiary dikes (22 Ma). In the Jizan area, the Al Birk lavas rest unconformably upon the Miocene Jabal Tirf dike swarm and on Wajid sandstone. On the coastal side, the Al Birk lavas overlie exhumed coral reefs that were formed along the former Red Sea shoreline. These stratigraphic relations indicate that the Al Birk harrat is Post-Miocene.

More than 200 cinder cones within the main volcanic field have a northerly trend rather than being parallel to the northwesterly trend of the Red Sea axial trough. These cones mark the main eruptive centers for individual flows. Many of the cones are now deeply dissected by erosion, but most of them were less than 150 m high with a basal diameter from 800 to 1000 m. The basalts are usually aphyric and uniform, with alkali olivine basalts predominating, a few of the flows are fractionated, and hawaiites and mugearites are present. Nearly all of the flows are undersaturated and show nepheline in the norm. The fluid flows generally moved towards the Red Sea coast. Locally, multiple flows formed barriers, forcing new lava to flow away from the coast. Xenoliths are common as fragments in cinder cones and within the solidified lava flows. In the Jizan area the inclusion suite is mostly harzburgite and dunite, with minor gabbro and fragments of Wajid sandstone. To the north the inclusions are harzburgite and dunite with lherzolite, gabbro, and websterite, accompanied by megacrysts of spinel, clinopyroxene, and plagioclase (Ghent et al., 1980; McGuire, 1988a).

Harrats Nawasif and Al Buqum (10,800 km²) form an elongate lava field situated east of Turba and southeast of Khurmah, forming a sloping plateau that has a maximum altitude of approximately 1400 m near its southern termination near the scarp and a minimum elevation of 1000 m near its northern termination, approximately 150 km northeasterly. Apparent flow directions are northeasterly away from the scarp. All of the flows rest on the Precambrian basement of the Hijaz Mountains and there is no evidence of interlayering with Tertiary sediments. North-trending faults within the Precambrian basement extend to the southern edge of the harrat and are truncated by the flows. The regional trend of the cinder cones, that mark the eruptive centers, is northeasterly.

The earliest flows, as much as 10 m thick, are exposed in the east draining, incised wadis. The main part of the harrat, a subtle northeast-trending highland, is a series of coalescing flows that emanate from more than 350 cones and small vents. The relative degree of cone erosion and the fresh appearance of some of the lavas suggest a northeasterly progression for eruptions, with the most recent being in the northeastern tip of the harrat. Most of the eruptions produced small cinder cones with a central vent. Some of the larger eruptive centers have small, shallow calderas. Alkali olivine basalts are the most common but there are also olivine transitional basalts present at the summits of some of the small cone flows. All of the basalts are undersaturated and nepheline normative, with two samples actually being nepheline basanites. The older flow surfaces contain numerous man-made stone stripes that may have been built during the period 3000 to 500 years B.C. Age determinations by Arno et al. (1980) and Brown et al. (1989) indicate that the oldest rocks are around 5 Ma, with volcanism continuing up to the recent past. Some older ages given by Arno et al. (1980) may be a result of excess argon, and the feeder neck reported by them to be ~23 Ma is probably related to Harrat Hadan, a few kilometers to the north.

Harrat al Kishb (6700 km^2) forms an elongate volcanic plateau the axis of which trends northerly. These lavas were extruded onto a peneplain of Precambrian rock and form a barrier to northward drainage from Sahl Rakbah. Sabkah deposits on the southeastern edge of Harrat al Kishb near Al Muwayh are a result of the lava flow barriers (Camp et al., 1992). Tertiary sediments are not found interlayered with the flows or below them. The underlying Precambrian basement has a northwesterly trend and is displaced by faults of the Najd system. However, the north trending vents and cinder cones within Harrat al Kishb do not appear to be controlled by exposed basement trends.

There are approximately 150 cinder and scoria cones within the harrat, and flows, usually less than 10 m thick, emanate from the cones and flow either to the east or west away from the central volcanic highland in the north of the harrat, which is dominated by two northward-trending eruptive centers. In the west the eruptive centers are comprised of shield volcanoes and pyroclastic cones, primarily consisting of alkali olivine basalt. To the east the eruptive centers consist of silicic tuff rings with central resurgent domes, sometimes breached by silicic flows that are mainly trachytes. The northern and southern parts of Harrat al Kishb are separated by the Holocene volcano, Jabal as Sauwahah. The alkali olivine basalt flows from this jabal abut the eastern margin of the Al Wahbah crater, 2 km in diameter and 270 m deep. The flat floor of the crater is covered by sabkah deposits and the vertical walls expose Precambrian andesites. Rimming the crater are basalt flows capped by a basaltic tuff ring bearing mantle xenoliths and megacrysts. Southeast of the crater is a very young obsidian flow that contains peridotite inclusions. Other Holocene activity is present in the northwestern corner of the volcanic highlands, where fissure eruption sent pahoehoe and aa lava westward and was terminated by pyroclastic activity which brought mantle-assemblage inclusions to the surface (Camp et al., 1992).

Eruptions of the harrat appear to have begun in the south and migrated northward with the most recent activity occurring in the central and northern parts of the harrat. Radiometric age determinations are not available but physiographic evidence indicates that volcanic activity began in the Pliocene–Pleistocene and continued up to the Holocene.

Harrat Rahat (18,100 km^2) extends southward from Al Medina to Wadi Fatimah near Makkah, a distance of 310 km with an average width of 60 km. The crest of the volcanic field is 650 m above sea level at Medina, rising to 1640 m in the central part. The southeast edge of the harrat forms the western barrier for Sahl Rakbah, where extensive sabkha deposits have formed. The high points of the crest are linear clusters of cinder cones, craters, and domes that represent four centers of eruptive activity coalescing to produce Harrat Rahat. From north to south these centers are referred to as Harrat Rashid, Harrat Bani Abdullah, Harrat Turrah, and Harrat Ar Rukhq. These centers have 20°N to 25°W trends each 50 to 75 km long and offset in a right lateral sense. Camp and Roobol (1989) have found four regional trends in Harrat Rahat: (1) north–south; (2) northwest to north–northwest (vent alignments); (3) east–west; (4) northeast. At least some of these trends coincide with the Precambrian basement trends. On the western margin of the harrat, numerous flows from the southern volcanic centers have breached passes in the Hijaz mountains and flowed westward onto the Red Sea coastal plain. The Shawahit flows consists mainly of olivine transitional basalt (90%) and minor alkali olivine basalt (10%); the Hammah unit consists of predominantly equal portions of alkali olivine basalt and hawaiite with very few domes and flows of mugearite and benmorite; the younger Medina basalts are quite variable, with alkali olivine basalts 47%, hawaiite 32%, olivine transitional basalts 8%, benmorite 8%, mugearite 4%, and trachyte 2% (Camp and Roobol, 1989).

On the west side of the harrat, erosion has cut gorges as much as 50 m deep through as many as 15 flows that had filled west flowing wadis. Some of these flows were interbedded with fanglomerates, indicating that there must have been considerable uplift of the area at the time of the early eruptions. In the west, harrat lavas rest on sediments of the Shumaysi formation of Oligocene age (Brown et al., 1989). Drilling 100 km south of Medina revealed marine sediments interlayered with the earliest flows, which have Miocene fossils (Durozoy, 1972).

FIG. 3.7. The Chada flow, which erupted 1256 A.D. The city of Medina is situated along the base of the distant hills. View looking northward along the edge of the flow from a helicopter at 250 m. The white areas are evaporite residues from ponded rain water (1981).

FIG. 3.8. Young basaltic collapse crater situated on the west side of Harrart Khaybar. Viewed from a helicopter about 100 m facing north (1981).

Studies on Harrat Rahat by Camp and Roobol, (1989) provide important constraints on the eruptive history. They divide the eruptive periods into three distinctive phases: (1) Shawahit basalt, 10 to 2.5 Ma (Brown et al., (1989), give maximum ages of 25 Ma); (2) Hammat basalt, 2.5 to 1.7 Ma; and (3) Medina basalt 1.7 Ma to present. According to Camp and Roobol (1989) the volcanism has migrated northward with time. The youngest historically recognized eruptive center is just east of Medina, where the Chada flow erupted about 1256 A.D. (Doughty, 1979; Camp et al., 1987) (Fig. 3.7).

Harrats Khaybar, Ithnayn, and Kura (21,400 km²) are coalesced harrats making up the largest continuous exposure of Tertiary volcanic rocks in Saudi Arabia. Harrat Khaybar rises gently toward the east from the main basalt plateau near the village of Khaybar. Along its northwest trending axis, several volcanic cones have elevations in excess of 1900 m.

The crest of the range contains two prominent intersecting trends: (1) a young, north trending chain of cones that includes basaltic shields, silicic tuff rings and domes, and intermediate composition tholoids; and (2) an older, N 25° W trending chain of basaltic pyroclastic cones (Fig. 3.8). These two belts contain several hundred volcanic centers. Jabal Abyad (white mountain), a prominent peak within the younger belt, is one of the largest and best preserved areas of silicic volcanic rock (Baker et al., 1973) (Fig. 3.9). This complex consists of tuff rings and domes that range in composition from mugearite to benmorite, trachyte, and comendite.

FIG. 3.9. Jabal Abyad (white mountain), a silicic eruptive dome surrounded by pyroclastic debris situated on the west side of Harrat Khaybar. View looking northward from helicopter at 320 m (1981).

Harrat Ithnayn is composed of a 40 km wide belt of volcanic centers aligned in northerly trends. The western chain of craters appears to be offset from the Khaybar trends in a right lateral sense. In Harrat Ithnayn, silicic rocks are absent but the pyroclastic cones contain numerous mantle xenoliths. Basalt of Ithnayn overlies Paleozoic sandstone on the southeastern margin of the Al Hisma plateau (Brown et al., 1989) and cascades over sandstone bluffs near the northern, eastern, and southern margins of the harrat. These relations suggest that the flows of Harrat Ithnayn are a thin veneer, no more than 100 m thick, covering the Paleozoic sandstone. Southwest of Harrat Ithnayn, the Al Hisma plateau terminates in a 300 m high erosional escarpment of Paleozoic sandstone, which marks the boundary between the Precambrian shield and the younger platform sediments. The escarpment is buried by the Khaybar lavas, suggesting that the lava pile may be 600 m thick. All of these lavas are thought to have been extruded onto basin-range topography because northwest trending ranges of Precambrian basement remnants interrupt the gently sloping surface of Harrat Khaybar.

Harrat Kura, to the west, is a sloping highland area that is deeply incised on its western margin. The basalt of the harrat forms a gently sloping surface whose dip varies between variable 2 and 5°. Deeply eroded pyroclastic cones are aligned N 30° W. The walls of these incised drainages expose a series of flows extruded on to the Precambrian basement, and as much as 100 m of relief. Younger flows from Harrats Khaybar and Ithnayn cover its eastern margin and partially fill the incised canyons of its northwestern margin.

Studies of the eruptive sequence in these three harrats by Camp et al. (1991) provide a consistent picture. The older Kura basalt ranges in age from 10 to 5 Ma and is predominately alkali olivine with only minor basanite and hawaiite These lavas are followed by the Jarad basalt (5 to 3 Ma) and the Mukrash basalt (3 to 1 Ma), which form the base of Harrat Khaybar and consist of mainly olivine transitional basalts. The Abyad basalt ranges from 1 Ma up to the present and is found mostly in the Khaybar and Ithnayn harrats consisting mainly of alkali olivine basalt and hawaiite with only minor olivine transitional basalt.

Harrat Hutaymah (900 km^2) is located 550 km east of the Red Sea and represents the most inboard harrat in Saudi Arabia. It consists of a thin veneer of basaltic rock in an area of low relief. It contains considerable basalt tephra along with minor amounts of basaltic flow rock. Tuff ring craters expose Precambrian basement under a cover of basalt flows less than 100 m thick (Fig. 3.10). The harrat consist of two separate, and north-trending eruptive trends. The basalts are all alkali olivine, forming flows fed from breached cinder cones of fissure vents overlain by tephra from the cones or tuff rings. The volume of erupted material in this area is quite small when compared to the other harrats. The western chain is dominated by explosive tuff rings and the eastern chain mainly pyroclastic cones dominated by basaltic tephra. Significantly these pyroclastics contain a wide variety of felsic and ultramafic xenoliths, representing material from the lower crust and mantle.

Fɪɢ. 3.10. Explosion crater within Harrat Hutaymah 550 km east of the Red Sea. The layered material exposed along the top edge of the crater consist of explosive base surge deposits, containing numerous mantle xenoliths and xenocrysts. View from helicopter at 120 m looking southeast (1981).

The lava flows overlay thin deposits of Quaternary sand and gravel, which in turn rest unconformably on the Precambrian basement. The preservation of the tephra cones and general preservation indicates a very young age for eruption, which is supported by a K/Ar age of 1.8 Ma (Pallister, 1984).

Harrat Lunayyir (1750 km²) is composed of older flows in the southwestern part and very young pyroclastic cones and flows to the northeast. It occupies the crest of the Red Sea scarp (Hijaz mountains), flanked by the Red Sea to the west and by Wadi al Hamd to the north and east. The crest of the range in the vicinity of Harrat Lunayyir is at 1500 m, decreasing to less than 400 m at the bottom of Wadi al Hamd. All of the flows postdate the formation and uplift of the Red Sea escarpment and extend in all directions from the range crest. Harrat al Qalib to the west is a single flow that extruded into the Red Sea. Alkali olivine basalts and hawaiites are the most common basalt types present.

The cinder cones show a rough northward trend, parallel to the magnetic anomalies reported by Blank, (1977). Kemp (1982) has shown that there were at least four separate eruptive events, and the older flows appear to be interlayered with Pleistocene corals on the coastline. K/Ar ages reported by various researchers on the youngest material range from 1 Ma to 0.5 Ma, so this eruptive center appears to be nearly equivalent to the Al Birk lavas some 800 km to the south.

Harrats Uwayrid and Ar Rahah (7150 km²) form a plateau nearly 230 km long with summit altitudes in excess of 1900 m in the central portion, decreasing to 550 m in the south near Al Ula. The two harrats are narrowly connected in the

FIG. 3.11. Western edge of Harrat Rahah where erosion has exposed feeder necks and dikes invading the Saq sandstones of Cambrian age. View looking northward 80 km east of the Red Sea (1981).

median portion, where numerous eroded feeder necks are present. Along the western edge of Harrat Rahah erosion has exposed domes and plugs connected by feeder dikes within the underlying Cambrian Saq sandstone (Fig. 3.11). The trends of the feeder dikes and the summit tephra cones is northwesterly parallel to the Red Sea axial trough.

In the western parts of the harrats, the lower lavas are extruded onto a tilted platform of Paleozoic sandstone and exhibit a slight angular disconformity, and the younger lavas flowed to the northeast away from the erosion scarp, suggesting that eruption was post-tilting. Brown et al. (1989) estimates a maximum thickness of 515 m with nearly 25 separate flows, gives K/Ar ages of ~9.5 Ma for the earliest flows, and suggests that the upper flows are Pleistocene to Holocene in age. The lower flows appear to be olivine transitional basalts that grade upward to olivine alkali basalts and finally to hawaiites. The most recent volcanic activity is in the Al Jaww Depression. Here the lavas flowed in a northeasterly direction but in other areas with inverted topography a northwesterly trending drainage on the west side indicates continued uplift during the final eruptions. The lavas of these two harrats are particularly rich in mantle xenoliths.

Harrat Shama forms a continuous volcanic field from Syria, across Jordan, and into Saudi Arabia. It extends nearly 500 km in a northwesterly direction varying in thickness from 100 m up to 1000 m in Syria. To the southwest, at Wadi Sirhan, drill holes penetrated basalt below as much as 200 m of sediment, and exposures along the wadi reveal basalts interbedded with shallow-water, marine sediments of Miocene age. Further north in Saudi Arabia the basal flows rest unconformably on lower Tertiary marine sediments 700 to 800 m above present sea level. The youngest volcanoes occupy the highest elevations, reaching a maximum height of 1800 m at Jabal Druse in Syria.

Approximately 500 volcanic centers lie along either older northwest-trending faults or younger north-trending faults. The older volcanic centers, found along the southeastern and southwestern margins, follow fault stuctures that trend northwesterly, whereas the younger centers south of Turayf follow north-trending structures. Van den Boom (1968) has divided the Jordan Harrat Shama sequence into six major flow units (B1–6). The older flows (B1–3) are not exposed at the surface but are considered to be Miocene in age. The B4 flows are 9 to 12 Ma and form shield volcanoes fed by a dike system. The B5 flows range in age from 7 to 9 Ma and form flood basalt flows. The youngest flows, B6, range in age from 0.1 to 0.8 Ma and cap the harrat with tephra cones, lava flows, shield volcanoes and feeder dike systems. Lowermost flows on the western side of the harrat near Hazawaza, Saudi Arabia, give K/Ar ages from 13 to 11 Ma (Upper Middle Miocene) and are probably representative of the older B1–3 flows of Van den Boom (Brown et al., 1989). Some of these eruptive centers are within the Jordan rift and have similar characteristics to the plateau basalts (Duffield et al., 1988).

For both the older and younger basalts, the predominant mode of eruption appears to have been through fissure, followed by pyroclastic cone building (Saffarini et al., 1985). Tuff rings are common in both the older margins of the harrat and in the younger interior south of Turayf. Most of the tuffs contain

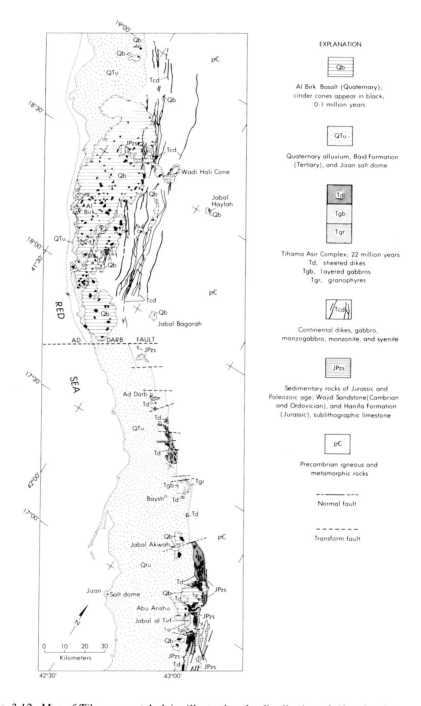

EXPLANATION

Qb

Al Birk Basalt (Quaternary);
cinder cones appear in black;
0·1 million years

QTu

Quaternary alluvium, Baid Formation
(Tertiary), and Jizan salt dome

Td

Tgb

Tgr

Tihama Asir Complex; 22 million years
Td, sheeted dikes
Tgb, layered gabbros
Tgr, granophyres

Tcd

Continental dikes, gabbro,
monzogabbro, monzonite, and syenite

JPzs

Sedimentary rocks of Jurassic and
Paleozoic age; Wajid Sandstone(Cambrian
and Ordovician), and Hanifa Formation
(Jurassic), sublithographic limestone

p€

Precambrian igneous and
metamorphic rocks

Normal fault

Transform fault

FIG. 3.12. Map of Tihama coastal plain, illustrating the distribution of rift volcanics and younger Al Birk alkali olivine basalts. From Coleman et al. (1977).

58

mantle xenoliths and megacrysts accompanied by inclusions of sedimentary rocks of the Cretaceous and Tertiary systems.

Rift Zone Igneous Rocks

A nearly continuous sheeted dike swarm can be traced almost 200 km from north Yemen to Ad Darb, Saudi Arabia, and discontinuous dike swarms are present in the Yemen Tihama to the most southern extension of the Yemen western escarpment (Capaldi et al., 1987b; Baldridge et al., 1991; Mohr, 1991) (Fig. 3.12). These dike swarms invade Precambrian basement and the overlying Phanerozic sediments along the northwesterly trend parallel to the Red Sea trough, and are most commonly found in the hinge line separating the coastal plain (Tihama) and the scarp area (Figs. 3.13 and 3.14). These dike swarms are terminated in the north by

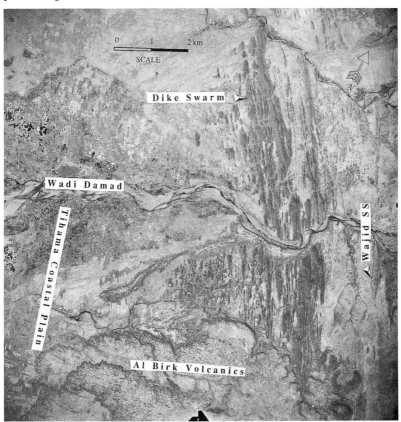

Fɪɢ. 3.13. Aerial photograph of Tihama dike swarm (vertical, black, parallel strips) just north of Jabal Tirf along Wadi Damad. Lobate flows of Al Birk basalt cover the dike swarm in the lower part of the photograph. Screens of Wajid sandstone (light colored) between dikes dip steeply westward. Some exposures along Wadi Damad reveal zones of spreading, where sheeted dike swarms intrude each other and have no visible Wajid sandstone country rock, similar to sheeted dikes found in ophiolites (oceanic crust).

F<small>IG</small>. 3.14. Tihama dike swarm along Wadi Bayd showing several generations of dikes. Where country rock could be established it was mainly older andesitic basalt. The larger dikes reach a maximum width of 1 to 1.5 m. Photograph from helicopter ~30 m above ground level.

the Ad Darb fault. Late Miocene post-emplacement listric faulting has rotated the dike complex toward the Red Sea axial trough, exposing the transition from older continental crust to newly formed mafic crust. At Jabal Tirf, this new crust consists of a 4 km wide zone of closely spaced, subparallel dikes with only minor screens of Precambrian schist and earlier rift volcanics between them (Schmidt et al., 1983) (Figs. 3.13 and 3.14).

Isolated, cone-shaped layered gabbro plutons within this dike complex represent small magma chambers developed during extension and concomitant rise of tholeiite magma (McGuire and Coleman, 1986) (Figs. 3.15 and 3.16). Granophyric sills and dikes along with shallow, irregular shaped plutons form complex intrusive relationships with mafic dikes and layered complexes. In some instances the granophyres intrude and cross-cut pre-existing mafic rocks, and in other cases mafic intrusives invade the granophyres (Fig. 3.17). These relationships indicate that both the granophyre and mafic magmas developed simultaneously. Mafic and granophyric dike widths range from 0.5 to 15 m, with the late cross-cutting mafic dikes typically being less than 1 m wide.

Extension north of the Ad Darb fault is manifested by much larger single dikes with widths up to 100 m. A single dike may be traced more than 50 km (Fig. 3.12). The individual dikes show a wide range in the degree of fractionation from gabbros to monzonites and quartz syenites, and their coarse-grained nature indicates slow cooling at hypabyssal depths. No eruptive equivalents of these dikes have yet been discovered and it is presumed that the magma density was greater than the surrounding crust, preventing eruption. It is possible that these dikes propagated horizontally as the magma reached a level where its density was equal to or greater than the enclosing crust. Blank (1977) has traced these dikes by utilizing their distinct elongate magnetic signatures on maps of the Arabian

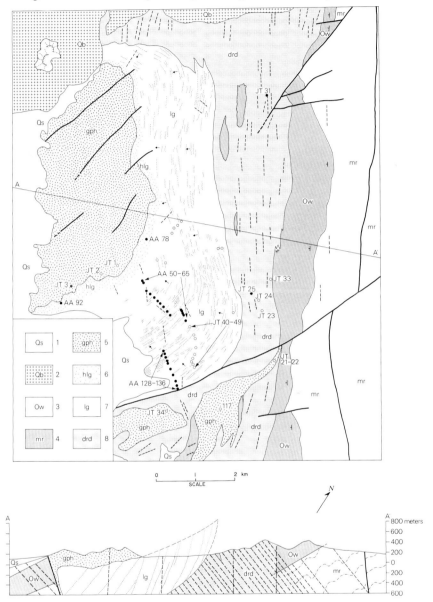

FIG. 3.15. Map of Jabal Tirf igneous complex illustrating mutual relationships between the layered gabbro, dikes, and granophyre. Explanation: 1, Quaternary coastal plain sediments; 2, Quaternary Al Birk Volcanics; 3, Paleozoic Wajid Sandstone; 4, Precambrian metamorphic rocks, Miocene Tihama Asir Complex; 5, granophyre; 6, high level gabrro; 7, layered gabbro; 8, diabase and rhyolite dikes, filled and open circles indicate sample locations from McGuire and Coleman (1986).

0   1   2   3 METERS

FIG. 3.16. Rhythmic layering exposed near the top of Jabal Tirf gabbro near locality
AA-78 in Fig. 3.15. Sharp dark contacts mark the bottom of phase contacts, where
cumulate olivine and clinopyroxene predominate over plagioclase. From McGuire and
Coleman (1986).

coastal plain, and was able to trace them to the Gulf of Aqaba. Strike-slip
movement on the Dead Sea rift has offset the dikes, and their northward extension
can be seen across the Sinai Peninsula parallel to the axis of the Gulf of Suez
graben (Bartov, 1980). Even though the dikes do not form dense conjugate sheets
a rough estimate shows that they may represent as much as 5% new mafic crust,
assuming they extend all the way through the crust. It is also possible that larger
amounts of new mafic crust have been underplated in the zone of extension along
the flanks of the Red Sea axial trough (Suayah et al., 1991). Reliable radiometric
ages of the individual large dikes, dike swarms, and gabbros are difficult to obtain
because of crustal assimilation of excess radiogenic argon from the surrounding
Precambrian rocks. Hornfels and rheomorphic zones along dike and gabbro
contacts yield ages of 22 to 25 Ma (Coleman et al., 1977).

The early extension of the Red Sea area was characterized by two distinct
types of tholeiitic dikes: (1) tholeiitic dike swarms forming sheeted sequences
and concomitant development of conical gabbro magma chambers, producing
zones of 100% new mafic crust; and (2) thick elongate individual dikes of
differentiated tholeiitic magma extending northward to the Sinai Peninsula.

Red Sea Axial Basalts

The narrow axial trough (5 to 30 km) is confined to the medial portions of the
basin. As mentioned earlier, the trough is marked by steep-sided walls and a very

(a)

(b)

FIG. 3.17. (a) Granophyre pluton of Jabal Tirf looking southwest towards the Red Sea and across the Tihama. The dark material at the base of the hill is the upper part of the layered gabbro (see Fig. 3.15). (b) Granophyre intruding earlier formed diabase dikes, as exposed in Wadi Jizan.

63

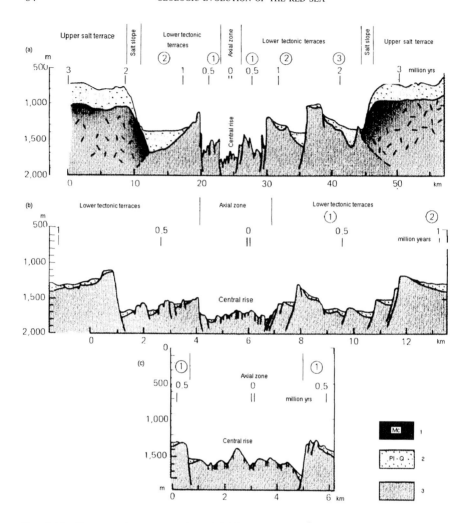

FIG. 3.18. Cross-sections across the axial rift zone near 18°N illustrating the structures within the spreading center. (a) Section across the entire rift (35 km) where salt deposits rest on the new crust covered by pelagic sediments containing Indian Ocean fossils. (b) Within the rift is a central rise (8 to 10 m wide) where young volcanoes erupt, and on the flanks marginal depressions are marked by fault blocks. (c) The inner axial zone is only 4 to 5 km wide and surrounded by fault blocks that rise 400 to 500 m above the axial zone (Monin et al., 1982). Explanation: 1, Miocene salt deposits; 2, Pleistocene-quaternary marine sediments; 3, axial basalts.

irregular bottom topography. Submersible dives in the deep trough have provided new facts concerning the evolution of these rocks (Zonenshayn et al., 1981; Monin et al., 1982). Volcanic activity can only be observed within the axial rift. Here it is possible to observe young fault scarps concentrated along the rift sides in zones of 10 to 12 km. The localization of volcanic and tectonic activity within this narrow zone is present all along the central portion of the Red Sea between latitudes 22° and 17°N (Fig. 3.18). In the northern part of the axial trough, extrusive volcanic areas are developed only where magnetic highs indicate subsurface intrusions such as sills and dikes. Where the axial trough narrows south of 16°N latitude, a shield volcano, Jabal Tair, of tholeiitic basalt rises 1200 m above the trough, inundating and covering the axial portions (Gass et al., 1973) (Fig. 3.19). Further south, the Zubayr volcanic island group consists of basalts intermediate in composition between tholeiites and alkali basalts, and some 300 km further south the Zukur–Hanish island group consists mainly of alkali-olivine basalts and leucocratic differentiates (MacFadyen, 1932; Gass et al., 1973). The presence of these alkali basalts in the axial part of the Red Sea indicates significant differences in magma types and may mark the most southern propagation of the axial trough basalts (MORB).

Fissures are abundant in the central zone, and form in clusters along the extrusive zone. They vary in width from less than 1 m to as much as 4 to 5 m, with a frequency of four or five fissures every 100 m of the rift cross-section. Zonenshain et al. (1981) estimates that these fissures constitute up to 6% of the internal rift width. The volcanic activity is characterized by fissure eruptions extruding on the inner floor within the deepest part of the trough. Small seamounts within the axial zone form conical shaped masses whose slopes are covered with lava pipes and rounded pillows (Fig. 3.20). Other conical shaped volcanic edifices within the axial trough and on its flanks are considered to be single seamounts. Some of the lower slopes contain abundant broken fragments of the same material.

FIG. 3.19. Jabal Tair, consisting of basaltic lava, forms a small island in the axial trough of the Red Sea at ~16° N. Darker strips on the surface are recent summit eruptions. Historic eruptions have been reported (Simkin et al., 1981).

FIG. 3.20. Pillow lavas from the inner axial zone. Picture taken from Pisces-XI working from the research vessel *Academik Mstislav Keldush* in 1979–1980, and provided by Dr. L. Zonenshain, Institute of Oceanology, Moscow.

The lack of sediment on these volcanic materials in the axial zone indicates a very young age, as the sedimentation rate in the Red Sea is high compared to deep abyssal oceans. In many places the lava tubes had broken open and were healed by vitrophyric (glassy) material, which forms numerous clots on the outer surfaces of the tubes.

Mapping of the linear magnetic anomalies within the rift zone has shown that the present spreading rate is 1.6 cm per year and for the last 2 to 3.5 Ma the spreading was 2 cm per year (Girdler and Styles, 1976). Therefore, it is estimated that the present axial trough formed within the last 5 Ma (Roeser, 1975). Older slumped deposits of evaporites covering the lavas on the flanks preclude surface sampling of the older mafic eruptions.

Xenoliths and the Red Sea Basin Mantle

Ultramafic inclusions from the mantle are common in the younger alkaline olivine basalts (AOB) in Saudi Arabia, Ethiopia, and Jordan. Locally, fragments of gabbro-granulite and Precambrian crustal rock are also present. These rocks provide information on the nature of the mantle and have been used to establish the thermobarometric regime of their mantle origin (Ottonello et al., 1980; Esperanca and Garfunkel, 1986; Kuo and Essene, 1986; Mittlefehldt, 1986; Nasir and Al-Fuqha, 1988; McGuire, 1988a,b; McGuire and Bohannon, 1989).

The majority of these inclusions are Cr-diopside harzburgites with some lherzolites and are considered to be representative of the mantle in this area. Pyroxenites that accompany these types are Cr-diopside websterites, spinel pyroxenites, and garnet-spinel websterites, and are thought to have formed from mafic magmas within the mantle. The relative abundances of these various kinds of xenoliths is variable within the Red Sea Basin, suggesting a heterogeneous mantle.

Using apparent equilibrium mineral pairs from these xenoliths it is possible to estimate a temperature and pressure history for inclusion suites (McGuire, 1988a,b) has shown that there is a regional thermal gradient in the mantle related to the present position of the Red Sea axial trough. For the Al Birk xenoliths near the coastline, temperatures of 1015 to 1040°C and pressures of 12 kbar were estimated, but several hundred kilometers inland temperatures and pressures for xenoliths from Harrats al Kishb and Hutaymah were 1000 to 1050°C and 13.5 to 16 kbar. These variations indicate a shallowing of mantle re-equilibration towards the Red Sea axial trough. A similar shallow mantle equilibrium for xenoliths from Assab, located on the margin of Ethiopia, shows temperatures of 950 to 1050°C and pressures of 10 to 8 kbar (Ottonello et al., 1980). Further inland Nasir (1992) reports that xenoliths from Aritain volcano give temperatures from 925 to 1025°C and pressures of 14 to 20 kbar. The Late Cretaceous Karem Maharal eruption in Israel gives yet another pressure/temperature fix on the undisturbed Precambrian mantle. Esperanca and Garfunkel (1986) and Mittlefehldt (1986), report temperatures from 990 to 1190°C with pressures of 16 to 30 kbar, consistent with the trend of deeper mantle sources away from the axial zone of the Red Sea.

McGuire (1988b) has pointed out that these pressure and temperature results reveal unusually high temperatures in the mantle beneath the Red Sea rift flanks, whereas surface heat flow measurements have values close to those expected for old Precambrian crust. This paradox of temperatures between the crust and mantle indicates a strong disequilibrium. McGuire (1988b), utilizing models of thermal re-equilibration of the lithosphere, concluded that the lithosphere thinning was initiated 15 to 20 Ma following the early rifting and volcanism. The mantle upwelling and concomitant uplift therefore seem to follow a passive rifting scenario brought about by plate movement and not by hot spot mantle convection.

The two pyroxene gabbro (granulite) xenoliths within some of the xenolith populations have generally retained cumulate textures and are thought to represent high-pressure mafic magmas underplating the extended continental margin (Ottonello et al., 1980; Coleman and McGuire, 1988). McGuire (1987) estimated temperatures of 830 to 980°C and pressures of 5 to 9 kbar, and suggests that these mafic rocks may represent older lower crustal material formed during the Pan-African event. Multiple origins for these mafic rocks seems entirely reasonable since some show granulitic metamorphic textures while others have undisturbed igneous cumulate textures. It is conceivable that the match between the lower crustal velocities and the granulites are indeed representative of the lower Precambrian crust. On the other hand, the large amounts of mafic underplating required under the presently extending continental margins can also be represented by these high-pressure and high-temperature gabbro xenoliths. Future coordinated studies of these xenoliths has the potential of yielding important controls on the thermal history of the mantle under the Red Sea Basin.

Petrogenesis

Four distinct magma environments can be recognized in association with the Red Sea Basin: (1) MORBs within the axial trough; (2) mafic dikes and layered gabbros invading extended continental margins; (3) granites and rhyolites invading extended continental margins; and (4) plateau basalts asymmetrically distributed on the Red Sea margins (Camp and Roobol, 1989; Camp et al., 1991, 1992).

The axial rift basalts in the Atlantis II deep are very close to the average MORB (Chase, 1969; Coleman et al., 1974; Coleman and McGuire, 1988). They contain phenocrysts of olivine, plagioclase, and clinopyroxene, with glomeroporphyritic clots of these same minerals in the glasses. Juteau et al. (1983) reports similar chemical parameters for basalts further south in the axial trough. However, these basalts are more-evolved ferrobasalts, characteristic of slow-spreading ridges.

The mafic, marginal, Miocene dikes formed in the early stages of spreading are generally hy and qz normative with plagioclase and pyroxene indicating their origin from a MORB-like parent. However, these dikes have all undergone strong fractionation at various depths within the attenuated crust (Pallister, 1987; Coleman and McGuire, 1988). The larger dikes north of Ad Darb that extend into the Sinai (Baldridge et al., 1991) follow low-pressure cotectics, resulting in quartz-bearing differentiates forming within the crust (Coleman and McGuire, 1988). A recent paper by Mohr (1991) details the extension of these dikes all along the Tihama to the south in Yemen. It has been suggested that at least some of these dikes exhibit fractionation trends developed under higher pressures within the lower crust (Bohannon, 1986a; Pallister, 1987).

Layered gabbros and their differentiates invading the extended continental crust have mineral compositions and characteristics similar to the Skaergaard gabbros of eastern Greenland, which also formed at shallow crustal levels during the early extension of the Atlantic (McGuire and Coleman, 1986). The presence of two-pyroxene gabbros as xenoliths within the basalts erupted through the extended continental crust suggests that at least some of the rising mafic magma was trapped in the lower crust–mantle interface, producing high-pressure magmatic assemblages (Coleman and McGuire, 1988; McGuire and Bohannon, 1989).

The plateau lavas of the Red Sea Basin show important petrologic temporal variations, and their various tectonic settings appear to control their compositions. The older harrats of Saudi Arabia (Sirat, Hadan, Ishara-Khirsat and Harairah) are generally more magnesium-rich olivine transitional basalts (OTB) in their early phases, and evolve towards undersaturated AOB, with nepheline appearing in normative calculations (Coleman and McGuire, 1988; Camp and Roobol, 1989; Camp et al., 1991, 1992; Du Bray et al., 1991). The Ethiopian plateau lavas are tholeiitic or OTB in the early eruptive stages and evolve towards AOB through several cycles (Mohr, 1971, 1983; Piccirillo et al., 1979). These Ethiopian plateau lavas also contain significant quantities of silicic lavas. The axial ranges within the Afar Depression have strong OTB characteristics but their K_2O content (~0.2%) precludes them from being classified as MORB (Mohr, 1983). The Yemen plateau lavas are a mixture of AOB and OTB interlayered with significant amounts of rhyolites and trachytes. It is not obvious from the data that there was any temporal progression from OTB to AOB (Chiesa et al., 1989; Menzies et al., 1990).

The younger plateau basalts (<5 Ma) are predominately undersaturated nepheline-normative AOB. Camp and Roobol (1989) and Camp et al. (1991) discovered that those plateau basalts erupting along a north–south axis called the Makkah–Medina–Nafud line (MMN) (Fig. I.2), passing through Harrats Rahat and Khaybar, are different from the other young basalts in having sequences that are predominately OTB with only mildly alkaline AOB. In a general way, all of these young plateau OTB eruptive sequences are followed by extensive AOB lavas that evolve to hawaiities. High-level magma chambers ponded in the upper crust may evolve even further to mugearites or benmorites, with extreme fractionation producing limited amounts of trachyte or comendite (Camp et al., 1991).

The Miocene–Oligocene A-type granites are closely associated with the mafic dike swarms (Mohr, 1991). The alkali granites of Yemen consist primarily of perthitic feldspar and quartz with some minor alkali amphiboles and acmite. These granites represent water-poor, hypersolvus magmas generated from parent alkali basalt magmas partially melting lower crust (Coleman et al., 1992). The granophyric, two-feldspar granites from Jabal Tirf are associated with mafic dike swarms and layered gabbros (McGuire and Coleman, 1986). These granites are formed by fractional crystallization from MORB-like magmas developed in the early stages of continental extension.

A comparison of these magma types using Harker diagrams ($K_2O + Na_2O$ versus SiO_2) produces strong contrasts (Fig. 3.21). The axial trough magmas of the Red Sea are consistently clustered around the average MORB composition whereas the earlier Miocene dikes are fractionated tholeiites with some OTB trends, which may be due to the hydrothermal introduction of alkalis. The underplated gabbros from the rift zones show a cumulate trend but clearly overlap the dike compositions. The A-type granites have a restricted high silica-alkal content and are considered to have been derived mainly by partial melting of the extended continental crust (Coleman et al., 1992). A-type granites in the Afar are of the same age and provide evidence of partial melting of the extended continental crust beneath the Afar volcanics.

Alkali–FeO*–MgO (AFM) plots of these same magmas illustrate a clear tholeiitic fractionation trend for the Miocene dikes and their close relationship with the more mafic underplated gabbros (Fig 3.22). The axial trough basalts

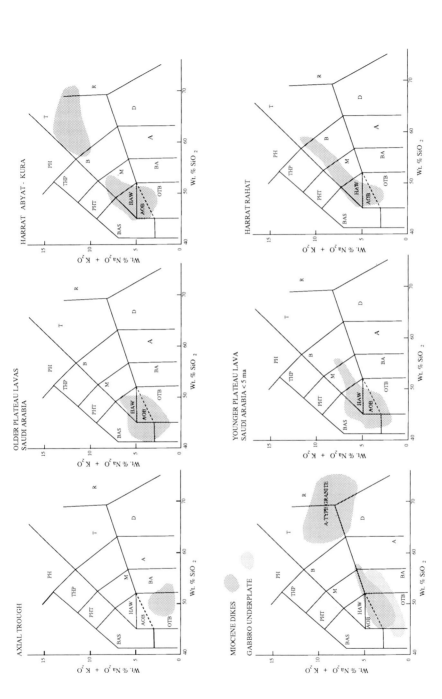

FIG. 3.21. Total alkali versus silica diagrams to illustrate variations in the magmas from Arabia. The classification used is that of Le Bas et al. (1986) and data are from Coleman et al., 1983; Coleman, 1984b; Coleman and McGuire, 1988; Camp and Roobol, 1989; Camp et al., 1987, 1991, 1992. Fields are as follows: BAS, basanite; OTB, olivine transitional basalts; AOB, alkali olivine basalts; HAW, hawaiite; PHT, phonotephrite; THP, tephriphonolite; PH, phonolite; T, trachyte; R, rhyolite; D, dacite; A, andesite; BA, basaltic andesite; M, mugearite; B, benmorite. The broken line

70

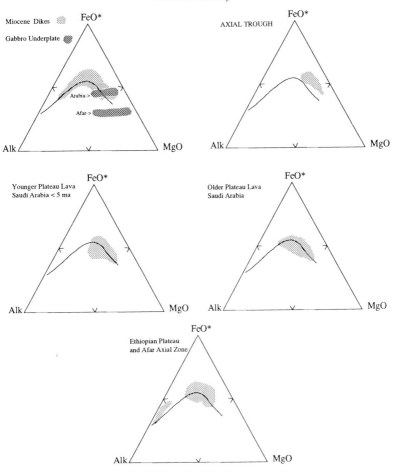

FIG. 3.22. AFM diagrams comparing trends for various magma types in and around the Red Sea. Data from sources quoted in the text.

show some fractionation but in general are almost exactly within the same field as MORB. The plateau basalts illustrate an ambiguous relationship with the primary AOB magmas plotted on the diagram. The older plateau basalts from Ethiopia and Arabia show abundant development of early OTB evolving into AOB, with higher level fractionation leading to hawaiites. The young basalts (<5 Ma) situated away from the MMN line in Arabia are mainly AOB with a few basanites fractionating to hawaiites and mugearites (Camp and Roobol, 1989; Camp et al., 1991). High level, restricted fractionation produces the trachyte-rhyolite with associated comendites. The axial basalts from the Afar are strongly fractionated with trends from OTB to basaltic andesite, trachyandestite to rhyolite, very similar to the trends found for the Miocene dikes on the Arabian margin. AFM plots for the plateau basalts illustrate the progression of the OTB to fractionate to

AOB. Primary melts from the mantle under deeper conditions of limited mantle melting and high pressures favor the formation of AOB, whereas the OTB are more common to rifting and result from larger scale melting at reduced pressures (Camp et al., 1991).

The axial trough basalts MORB contain (<1.5%) TiO_2, and with fractionation of MORB magma forming the Miocene dikes TiO_2 increases up to 3%, still much less than the 4% TiO_2 present in the AOB. General comparisons made by other researchers bear out the observation that the OTB and AOB are much more strongly enriched in the incompatible elements when compared to MORB mainly because of the differences in the degree of partial melting.

Chondrite-normalized rare earth elements (REE) patterns provide further evidence for the distinctness of the Red Sea magmas (Fig. 3.23). The axial zone MORB have REE concentrations ×10 chondrite or less, and generally form flat curves. Basalts from the southern Red Sea show a depletion in light REE. Altherr et al. (1988) have shown a non-systematic variation of REE content in samples from different segments of the Red Sea axial trough. Samples from the Zubayr and Hanish islands and from Jabal Tair have strong enrichment in the light rare earth elements (LREE) whereas samples from the central Red Sea (Atlantis II and Shagra Deep) have strong depletions of LREE, characteristic of N-type MORB. Other samples from these areas taken by Altherr et al. (1988) have normal MORB REE patterns, so there does not seem to be any evident systematic variation of the REE contents of the axial trough basalts. Much more detailed sampling is required to understand local variations before large-scale generalizations can be made. The Miocene dikes have enriched patterns of REE nearly ×30 chondrites but these patterns are almost parallel to the axial trough MORB REE patterns, indicating a progessive enrichment by fractionation. Enrichment of the LREE of the Miocene dikes suggests that there has been some contamination by assimilating extended continental crust. The strong enrichment of the total REE in the granophyres over the Miocene dikes reflects both differentiation and crustal assimilation (Coleman et al., 1992).

In contrast to the rift-generated tholeiitic basalt magmas, REE data for the oldest Arabian plateau basalts show distinctly different patterns of fractionation. The earliest flows are OTB and the least fractionated of these has a ×2-5 enrichment of the LREE and medium REE (MREE) when compared to the MORB patterns. However the heavy REE (HREE) are nearly equal to the axial trough MORB. With decreasing age the older harrats have strong enrichment of LREE and MREE, with patterns typical for other AOB (Wilson, 1989). The older Ethiopian plateau lavas show a similar transition from OTB to AOB with increasing LREE and MREE (Hart et al., 1989). Camp and Roobol (1989) and Camp et al. (1991) provide REE data for the younger (<5 Ma) Arabian plateau basalt series. The earlier OTB eruptions show moderate enrichment in LREE and MREE whereas the later AOB show strong enrichment of LREE and MREE, with the HREE nearly constant for both the OTB and AOB. Camp et al. (1991) has further emphasized that the basalts erupting along the MMN line are generally less enriched in the LREE and MREE which may be interpreted as indicating a higher degree of partial melting.

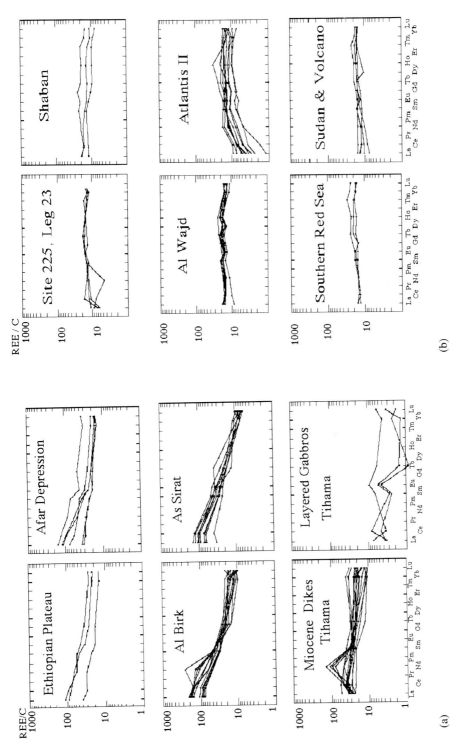

FIG. 3.23. REE patterns of various magma types in and around the Red Sea. (a) Saudi Arabia, Ethiopia, and Afar comparisons. (b) Axial trough basalt comparisons. Data are from Coleman et al. (1983), Betton and Civetta (1984), Altherr et al. (1988), Coleman and McGuire (1988), Hart et al. (1989), compared to values of Nakamura (1974).

73

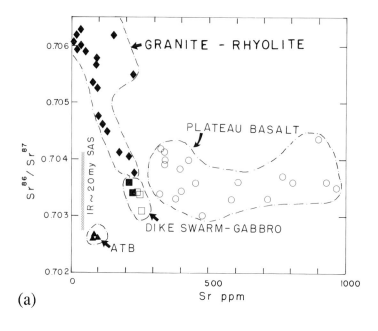

(a)

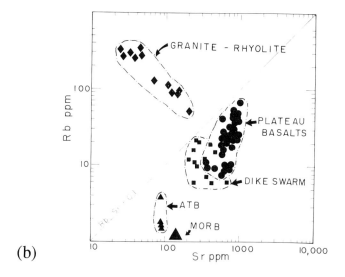

(b)

Fig. 3.24. (a) Rubidium–strontium plot. (b) $^{86}Sr/^{87}Sr$ to Sr concentrations. (▲, ATB), Axial trough basalts; (■), Miocene dikes; (□), Miocene gabbros; (♦), A-type granite-rhyolite; (○, ●), plateau basalts (AOB, OTB). Initial ratio (IR0) at 20 Ma for the Saudi Arabian shield (SAS) is shown as a vertical line. From Coleman and McGuire (1988).

Numerous authors have suggested that the lavas of axial zones within the Afar Depression are probably MORB-like. However, the REE patterns of these rocks are very similar to the OTB and show evidence of crustal contamination.

These REE data provide additional reinforcement of the theory that Red Sea magmas are derived from diverse sources and represent magmas generated by unique episodes of partial melting in the mantle.

The Red Sea magma types exhibit strong variations in their rubidium and strontium contents. The Miocene dike swarms and the plateau AOB and OTB are usually enriched in both rubidium and strontium when contrasted with the axial trough MORB. These higher values in the AOB and OTB are probably the result of initial small melt volumes, whereas the Miocene dikes with higher rubidium/strontium content may result from crustal contamination during their intrusion into the extending crust. The differentiated rocks from the Miocene rift zone, as well those associated with plateau lavas, have extremely low strontium content (<100 ppm) increasing the rubidium/strontium ratio to greater than 1.

Initial ratios (IR) of $^{87}Sr/^{86}Sr$ to Sr for the main magma types are shown in Fig. 3.24. It is apparent that the IRs for AOB and OTB are consistent and range from 0.703 to 0.704, with Sr usually exceeding 300 ppm. The Red Sea MORB axial trough basalts have the lowest $^{87}Sr/^{86}Sr$ IR and correspond to typical mantle values found for most MORB. The Miocene dikes and cogenetic A-type granites show a significant shift to higher IR with Rb/Sr≈8. These data indicate that the

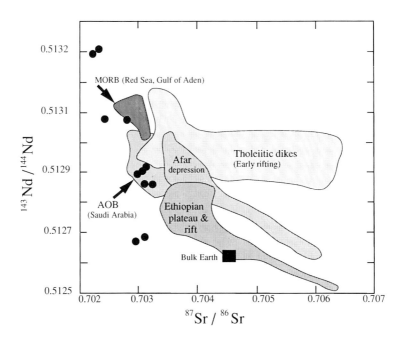

FIG. 3.25. Neodymium–strontium summary diagram. Closed circles represent mantle minerals and peridotites from the Red Sea area. Data are from Zindler and Hart (1986), Hart et al. (1989), Altherr et al. (1990), Schilling et al. (1992).

AOB assimilated very little crustal material on their way through the crust. The high strontium concentrations of the AOB and OTB precludes significant contamination by the older Precambrian crust, which contains less than 30 ppm strontium, with an estimated Tertiary IR between 0.703 and 0.704 (Fleck and Hadley, 1982). The Afar OTB ranging between 0.703 to 0.704 is higher than the Red Sea axial trough MORB, with strontium <300 ppm (Barberi et al., 1980).

Distinctions can be made between magma types by using a $^{87}Sr/^{86}Sr$ versus $^{143}Nd/^{144}Nd$ correlation diagram (Zindler and Hart, 1986; Wilson, 1989) (Fig. 3.25). A number of separate measurements have been made on neodymium and strontium isotopes on the Red Sea magma types and it is possible to suggest sources and contaminants (Betton and Civetta, 1984; Stein et al., 1987; Brueckner et al., 1988; Hegner and Pallister, 1989; Altherr et al., 1990; Barrat et al., 1990). The MORB field represents oceanic spreading centers and is depleted relative to the bulk earth composition, which is a time integrated value (Allegre, 1987). The extraction of material from the mantle leaves a residue that becomes depleted in rubidium relative to strontium, and in neodymium over samarium. Large-scale, shallow partial melting of depleted mantle produces MORB. The Red Sea and Gulf of Aden axial trough basalts all plot within the MORB field, indicating their formation within an ocean spreading ridge unaffected by adjacent continental crust. The AOB and OTB plateau basalts of Arabia form a grouping displaced from the MORB, suggesting a unique mantle source unrelated to the oceanic spreading centers. Peridotites and minerals from Zabargad Island, as well as xenocrysts, plot to the left of the mantle array overlapping with the plateau AOB–OTB, indicating a similar mantle source (Brueckner et al., 1988). Since the age of the Red Sea MORB and the plateau AOB–OTB are similar, it appears that these magmas formed in distinct regimes within the mantle. The early Miocene dikes show strong enrichment in $^{87}Sr/^{86}Sr$ but they overlap both the MORB and AOB plateau fields, indicating an origin from MORB-like magma (Hegner and Pallister, 1989). Hydrothermal circulation during and after consolidation by seawater has modified the initial strontium ratios in the Miocene dikes. The Afar basalts are displaced from the mantle array and both their REE contents and neodymium-strontium isotopes indicate large-scale continental contamination, as well as derivation from a less depleted mantle. The Ethiopian plateau lavas follow a similar trend but have experienced a much higher degree of contamination and fractionation.

The presence of a hot spot in the central portion of the Afar has been suggested by Schilling (1969, 1973) and Schilling et al. (1992). It is not clear how the hot spot magma will change over time to accommodate the eruptive sequences found in the Afar. It is possible that the early Afar eruptions could have been derived from a source known as prevalent mantle (PREMA) (Carlson, 1984). All of the later eruptive sequences appear to have PREMA affinities, with modification by fractionation and contamination. The MORB lavas are from strongly depleted mantle (DMM), and clearly follow punctiform expansion along the Red Sea and the Gulf of Aden unrelated to the Afar hot spot (Bonatti, 1985). The AOB plateau basalts require a special enriched mantle undergoing limited partial melting. The wide geographic distribution of the AOB and OTB in Arabia and their diachronous eruptions precludes them from being derived from the Afar hot spot PREMA magmas. Camp et al. (1991) make a good case relating the AOB–OTB plateau lavas to the Hail Arch or MMN line, an antiform that contains mainly OTB on its crest with AOB along the flanks, indicating mantle upwelling along the crest. The Miocene dikes produced in the early rifting have REE and neodymium and strontium isotopic signatures that indicate an initial widespread development of MORB-like magmas filling and underplating extended continental crust from the Sinai to Yemen unrelated to the Afar hot spot. Late hydrothermal circulation of hydrothermal fluids has greatly enriched these rocks in ^{87}Sr, similar to that found in the Oman ophiolites (McCulloch et al., 1981).

Constraints Developed from Volcanic Rocks

1. The variety of magma types indicates multiple mantle sources.

2. The volume of the alkali plateau basalts is much less than the volume of the axial trough MORB and related gabbro underplating.

3. Plateau basalt eruptive centers change from OTB to AOB through time, becoming increasingly undersaturated due to smaller amounts of partial melting in the mantle.

4. Thermobarometric studies of xenoliths indicate that mantle convection under the Arabian plate is recent and not in equilibrium with the upper crust.

5. The magma systems developed since the Early Miocene initially responded to passive rifting rather than a central hot spot, and were erupted near sea level rather than on the crest of a thermal dome.

6. Post-eruption uplift in the Red Sea appears to be mainly connected to underplating along the flanks of the Red Sea Basin.

4

Age Relationships

The evidence so far accumulated on the timing of volcanism and the evolution of the Red Sea reveals an extremely consistent pattern of kinematic movements and creation of magmatic rocks within the Red Sea environs (Figs. 4.1 and 4.2). It is generally accepted that some sort of faulting or extension is required to initiate volcanism, so establishing the ages of the volcanic events in and around the Red Sea provides insights to this relationship.

Early Stages

The basal portions of the Ethiopian plateau lavas contain eruptive sequences (Ashangi formation) that represent the oldest basaltic eruptions in the region. These are generally considered to be of the Eocene period, although there is considerable controversy about the exact span of eruption and correlation over vast areas of Ethiopia (Jones, 1976; Piccirillo et al., 1979; Davidson and Rex, 1980; Zanettin et al., 1980a,b; Mohr, 1983; Bohannon, 1986b; Menzies et al., 1990). Unconvincing evidence of Eocene volcanic and intrusive activity in Arabia is provided by Pallister (1987) and Menzies et al. (1990). A single flow in southeast Yemen resting on the Cretaceous Tawilah sediments near Rada, gives a K/Ar age of 43.5 Ma (Menzies et al., 1990), and a separated plagioclase from a hawaiite dike from the Damn complex near Al Lith gives a K/Ar age of 43.5 Ma (Pallister, 1987). Assuming these ages to be correct, these Middle Eocene mafic magmas could signal the earliest phases of extension. However, there is no structural evidence of regional faulting at this time.

Widespread igneous activity in the Arabian–Nubian region beginning in the Late Oligocene about 30 Ma and extending to the Middle Miocene about 15 Ma marks the early phases of rifting in the Red Sea and Afar. On the Ethiopian plateau the Aiba basalts produced a massive outpouring in the western plateau (32 to 28.5 Ma). Rhyolitic ignimbrites appeared during the Late Oligocene (28 Ma) in a diachronous way, intercalating with the plateau lavas, and continued into the Late Miocene. Small plutons of A-type granite on the flanks of the plateau in the Early Miocene are further evidence of either large-scale crustal melting or differentiation. In Yemen the basal basalt flows of the Trap Series are Late Oligocene (30 to 28 Ma), with two distinct periods of eruption between 30 to 26 and 23 to 19 Ma. In some areas they rest on a lateritic paleosol developed on the Tawilah formation (Chiesa et al., 1989). Rhyolites and rhyolitic ignimbrites are

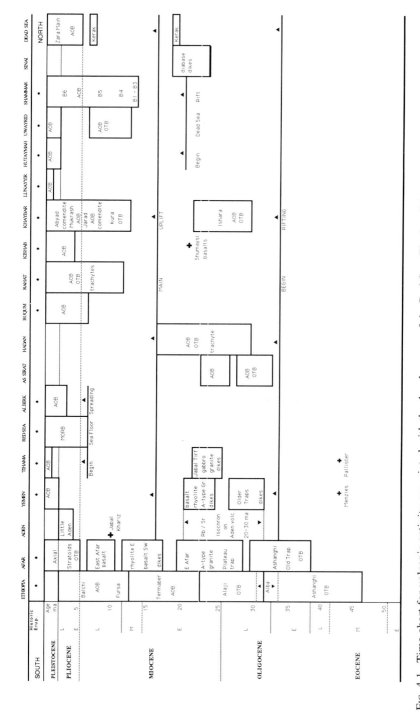

FIG. 4.1. Time chart for volcanic activity associated with the development of the Red Sea. The localities progress from south to north (left to right). Nearly all of the ages are K/Ar determinations from various sources, as discussed in the text.

intercalated throughout the eruptive sequence, accompanied by hypabyssal A-type granites at around 25 Ma (Capaldi et al., 1987a). Tholeiitic diabase dike swarms along the base of the scarp represent new mafic crust in zones 5 to 10 km wide (Mohr, 1991) (Fig. 4.1).

RED SEA BASIN

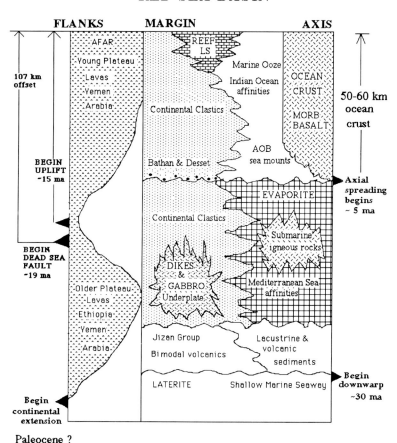

FIG. 4.2. Tectonic-stratigraphic progression of the formation of the Red Sea Basin based on volcanic, stratigraphic and structural data from numerous geologic sources. Timing of the Dead Sea fault is based on K/Ar ages of dike swarms (Coleman et al., 1977; Bartov, 1980). The age of the Jizan group is based on paleontological evidence and K/Ar ages (Schmidt et al., 1983). Continental extension is considered to start in the Late Oligocene, and is marked by the beginning of volcanism and block faulting. Zones of dikes and gabbros are defined from discontinuous screens of new crust in the attenuated continental crust, as typified by the Tihama Asir complex. Ocean crust formation and spreading is marked by the major unconformity above the evaporites at about 2 to 5 Ma, when MORB form new crust in the axial trough along with AOB sea mounts at the southern end of the Red Sea propagating rift.

Just north of the Yemen border, the basal flows of the Sirat volcanic field yield Late Oligocene K/Ar ages (31 Ma) and rest on a lateritic surface similar to that found in Ethiopia and Yemen. The upper flows give Early Miocene ages (22 Ma). The horizontal Sirat flows are 1 km thick and are now at an elevation of 3000 m (Du Bray et al., 1991). The tholeiitic diabase dike swarm present in Yemen at the base of the scarp extends northward along the Arabian Tihama and at Jabal Tirf an Early Miocene (20 to 24 Ma) hypabyssal gabbro-granophyre complex intrudes the dike swarm (Coleman, 1984b). The dike swarm changes at the Ad Darb fault (see Fig. 3.12), and coarse-grained thick dikes extending to the Sinai are also Early Miocene in age (Blank, 1977; Steinitz et al., 1978; Coleman, 1984b; Pallister, 1987). The best-documented stratigraphic relationships of this Late Oligocene–Early Miocene event are found at Harrat Hadan east of Taif (Madden et al., 1980; Brown et al., 1989). These lavas, consisting of brackish water esturine deposits containing a rich vertebrate and invertebrate fauna indicative of a Late Danian or Early Thanetian age (Paleocene) rest on latertized paleosol above the Umm Himar formation. Small remnants of the Late Oligocene–Early Miocene igneous event can be seen just north of Medina, nearly 1100 km north of the Afar triple junction. Harrats Ishara-Khirsat and Harairah are deeply dissected and eroded remnant lava plateauwhose K/Ar ages range from 28.3 to 21.2 Ma. The basal flows rest on Post-Eocene alluvial gravels containing chert boulders with clasts of Eocene marine invertebrate fossils (Brown et al., 1989). The Shumaysi formation, a clastic ironstone produced by erosion of the Paleocene laterite, contains interlayered basalt (20.1 Ma K/Ar) (Coleman et al., 1979) near the top of the section where Oligocene–Miocene invertebrate fossils have been identified (Brown et al., 1989). These radiometric ages and stratigraphic relations demonstrate a widespread period of volcanic activity. The laterite of As Sirat and Yemen indicates that this area was near sea level prior to initiation of volcanism, and basal marls at Harrat Hadan indicate that it was just at sea level prior to eruption. Marine sediments of the Harrat Hadan area may extend northward to Turayf and into Jordan (Harrat Shama). Isolated outcrops of sedimentary rocks containing marine fossils, in some places overlain by alluvial clastics, suggest that a shallow, narrow, fault-bounded seaway extended southward from Turayf. The presence of both marine and coarse alluvial deposits beneath the basal basalts of Harrats Harairah, Ishara-Khirsat, and Hadan may have been controlled by north–south extensional faulting, producing lava-filled basins which now show topographic inversions.

These early eruptive events from Ethiopia, northward to Yemen and to Saudi Arabia, and on to Jordan, were widespread and seem to be related to a continental rifting event that was initiated in the Late Oligocene and extended sporadically up to the Middle Miocene. Following this initial tectonic–magmatic phase during the Late Oligocene and Early Miocene, volcanism waned in the Red Sea and Afar basins in the Middle Miocene (16 to 14 Ma). During this time left-lateral movement along the Dead Sea rift zone indicated a change in the kinetic pattern of plate movement (Bartov, 1980) and extensional faulting on the eastern and western margins developed on the Afar margins (Mohr, 1978). On the Arabian plate AOB extrusions commenced along north-trending fractures at Harrat Rahat (10 Ma),

Harrat Khaybar (11 Ma), Harrat Uwayrid (12 Ma), and Harrat Shama (13.4 Ma), and along the Dead Sea rift (Sebai et al., 1987, 1991; Camp and Roobol, 1989; Camp et al., 1991, 1992). In the Afar, dike swarms developed on the margins (15 to 11 Ma) (Mohr, 1978). In the Gulf of Aden, new ocean crust began to develop around 10 Ma and continues up to the present time (Laughton, 1966; Cochran, 1981).

Late Stages

These Middle Miocene eruptive centers on the Arabian plate and Afar Depression have nearly continuous eruptive sequences up through the Holocene. At the beginning of the Pliocene (about 5 Ma), ocean crust began to form in the axial portions of the Red Sea between latitudes 22° and 15°N, and new eruptive centers formed at Harrat Al Birk (2.61 Ma), Harrat Buqum (5 Ma), Harrat Khaybar (5 Ma), Harrat Hutaymah (1.8 Ma), Harrat Ithnayn (3 Ma), Harrat Kishab (2.4 Ma), Harrat Lunayyir (1.4 Ma), Harrat Uwayrid (1.4 Ma), and Harrat Shama (5 Ma), and at the Dead Sea rift in Israel and Jordan at Karak (6 Ma) and Zara Main (3.7 Ma) (Siedner and Horowitz, 1974; Coleman et al., 1977; Brown et al., 1989; Camp and Roobol, 1989; Nasir, 1990; Camp et al., 1991, 1992). On the very southern tip of the Arabian plate, the Aden volcanics formed between 6.4 and 5.0 Ma (Cox et al., 1969, 1970; Dickinson et al., 1969) and minor Holocene eruptions are found in the Yemen highlands. All of these areas have evidence of Holocene activity, but the volume of these eruptions is much less than that of the earlier Oligocene–Miocene events. In the Afar, the Stratoid basalts form extensive flows over most of the basin, beginning at 4.3 Ma and terminating at 0.6 Ma. Eruptions along the axial range commenced about 1.2 Ma and continue up to the present (Barberi et al., 1972; Mohr, 1978) (Fig. 4.1).

These constraints on the igneous timing provide clear guidelines for developing an overall scheme of volcanism and plate movements. Although minor igneous events can be traced as far back as the Middle Eocene, the age data indicate that the Red Sea area experienced its first major continental rifting and concomitant igneous activity in the Late Oligocene (about 30 Ma). During the Late Oligocene to Early Miocene (15 to 30 Ma) the Afar Depression formed and, at the same time, the Red Sea depression began to form by extension and development of new mafic crust by underplating and intrusion of tholeiitic magmas in dikes. Uplift of the Red Sea margins began in the Middle Miocene (13.8 Ma) and relates to the earlier rifting brought on by asthenospheric uprising (Bohannon et al., 1989). As the Gulf of Aden opened in the Late Miocene (10 Ma), renewed AOB formed on the Arabian plateau (5 to 13 Ma) and at the beginning of the Pliocene MORB basalts developed in the Red Sea axial trough (Fig. 4.1). Renewed eruptions of transitional basalts in the Afar indicate nearly complete thinning of the area and formation of new mafic crust, but not yet an oceanic spreading center, as formed in the Red Sea. Continued eruptive activity in the Afar Depression along the axial range, within the Red Sea axial trough, and on the Arabian plateau is evidence of sustained plate motion, and mantle convection continues up to the present (Fig. 4.2).

5

Red Sea Structure

The structural aspects of the Red Sea basin are difficult to systematize because of the diverse structural styles and variability of the rocks exposed, in an area more than 2000 km in length (Fig. I.2). The Precambrian basement rocks are present in many areas and because of their metamorphism, plastic deformation and lack of coherent stratigraphy it is difficult to establish Tertiary timing of faulting or folding. With new ideas of continental marginal extension, the older simpler idea (Wegener, 1924; Arkell, 1951; Picard, 1970) that the Red Sea consisted of a large master graben has been abandoned. This chapter will focus on the areas where newer and more complete structural information has been developed.

Pre-Rift Configuration

There have been numerous attempts to reconstruct the Arabian shield across the Red Sea Basin, the most successful of which is that of Vail (1985), incorporating all of the known data for Saudi Arabia, Egypt, Sudan, and Ethiopia (Fig. 5.1). The most important aspect of this reconstruction is the recognition that the shield rocks are a collage of separate terrains, that representing a series of accretions that led to continental accretion (Camp, 1984; Stosser and Camp, 1985; Pallister et al., 1987). Older Proterozoic crust is generally lacking in this area but on the most western and eastern parts older gneissic terranes are present (1700 to 1185 Ma). These older terranes usually consist of quartzo-felspathic gneisses, and migmatites with metamorphic grades of amphibolite to granulite facies. Nearly all of these rocks carry isotopic signatures that indicate older continental cratonal crust (Stacey and Hedge, 1984). In between these older cratons, approximately six separate terranes separated by mobile belts are recognized. These terranes are made up of predominantly volcano-sedimentary rocks, mainly andesitic tuffs and lavas associated with minor rhyolites. These units are nearly always overprinted by greenschist to low amphibolite facies metamorphism. Geochemical studies of these rocks have defined them as island arc or plate margin assemblages. Numerous isotopic age determinations reveal that they formed between 800 and 500 Ma and that their lead isotope signatures all point to derivation from igneous material having island arc and oceanic signatures (Camp, 1984; Stosser and Camp, 1985; Pallister et al., 1987).

    The more remarkable aspect of these terranes is that they are often separated by ophiolite belts whose primary ages are all older than the arc terranes they

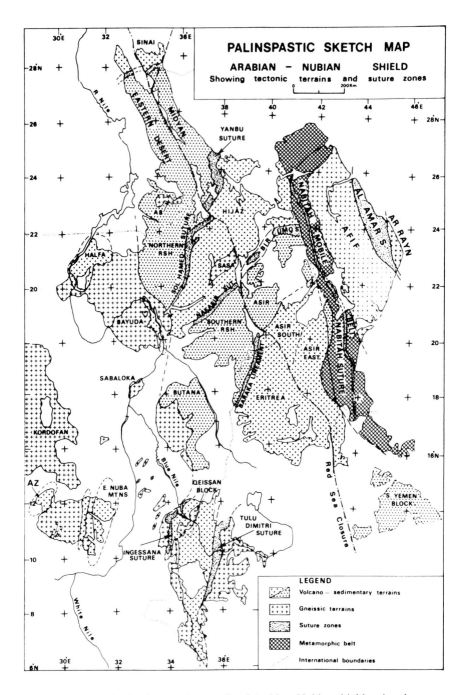

FIG. 5.1. Palinspastic sketch map showing fit of Arabian–Nubian shield, using the concept of terrains and reconstructed suture zones. From Vail (1985).

separate. Ophiolite ages are around 900 Ma, indicating active ocean spreading at the end of the Proterozoic (Pallister et al., 1987). Many researchers have interpreted the ophiolite belts as major suture zones between inter-oceanic plates represented by the terrane collage. It is interesting to note here that these ophiolite belts are more likely to represent continental growth and accretion of oceanic crust along active continental margins, rather than marking major collisional sutures (Stosser and Camp, 1985; Vail, 1985; Pallister et al., 1987). Attempts to show that the major sutures may have been the main factor in localizing the initial rifting of the Red Sea have not been convincing, because the Red Sea trend mainly cuts across these sutures at fairly high angles (Stern et al., 1984; Stosser and Camp, 1985; Vail, 1985; Dixon et al., 1987, 1989). The zigzag patterns found along the margins and within the Red Sea undoubtedly demonstrate local control on the distribution of extension in the upper crust, but there seems to have been an earlier strike-slip sinistral motion structure that may have localized the Red Sea in its present position (Shimron, 1990; Makris and Rihm, 1991). Perhaps the overall continental thickness was less in the area of the Red Sea as a result of incomplete continental thickening. Another factor important to this discussion is the Pan-African thermal trend, which in general follows the Red Sea trend and may therefore represent an earlier signature of a large elongate mantle plume now controlling the opening of the Red Sea.

Gulf of Suez

The Gulf of Suez is considered to be an asymmetric graben nearly 300 km long (Robson, 1971; Garfunkel and Bartov, 1977; Garfunkel, 1988; Lyberis, 1988) (Fig. 5.2). The actual structural width of the basin is about 60 km using the uplifted Precambrian outcrops on the Egyptian and Sinai sides (Ott d'Estevou et al., 1987). The Gulf of Suez has been divided into three separate basins representing half grabens. The average tilt of the sediments is about 20° but is reversed across northeast–southwest fault zones that mark the boundaries of the basins (Colleta et al., 1988). Fault patterns developed during rifting are partly controlled by inherited basement structures (Greiling et al., 1988; Montenant et al., 1988; Perry and Schamel, 1990) and follow a general zigzag pattern. Normal faults developed parallel to the elongation of the Gulf of Suez during rifting and, in the early stages, tilted blocks formed and shed debris into the smaller basins (Montenant et al., 1988). Later movements produced horst and graben structures related to subsidence (Steckler et al., 1988) while later peripheral uplift elevated the shoulders of the basin (Montenant et al., 1988). No direct evidence of regional doming around the Suez basin prior to rifting has been established either by structural evidence or sediments related to uplift (Steckler, 1985; Evans, 1988).

Gulf of Aqaba

The Gulf of Aqaba (Elat) is the southern end of the Dead Sea rift. It forms a deep trough 180 km long and is considered to be a transform boundary terminating the northern Red Sea spreading centers (Quennell, 1956, 1958; Ben-Avraham et al.,

1979b; Bartov, 1980; Mart, 1982; Ben-Avraham, 1985, (Fig. 5.2). The Gulf of
Aqaba is controlled by faults that are considered to have predominantly a left
lateral motion (Ben-Avraham et al., 1979b). The exposed faults that border the
Gulf of Aqaba have nearly vertical and very straight scarps. These faults are
subparallel along the west boundary of the Gulf and show mainly sinistral

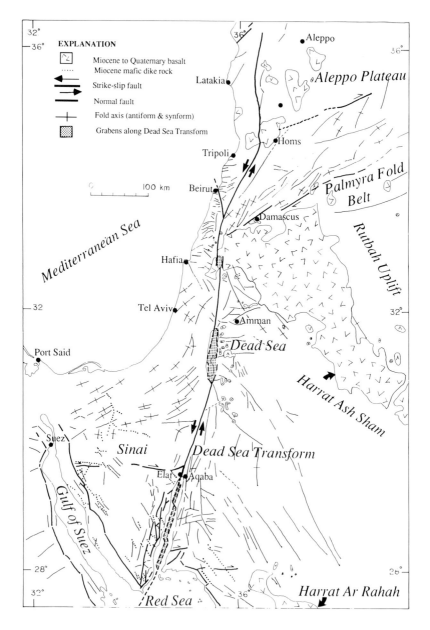

FIG. 5.2. Structural outline map of the Levant–Dead Sea–Gulf of Suez system.

displacements up to several kilometers (Bartov, 1980). Oceanographic studies of the Gulf of Aqaba have shown that active normal en echlon faults border elongate depositional basins (Ben-Avraham et al., 1979a). The offset of mafic dikes (19 to 22 Ma, representing magmatic activity resulting from intial rifting of the Red Sea) is used to fix the final stages of the opening of the Red Sea (Bartov, 1980). Faulting and rapid clastic sedimentation within the Gulf of Aqaba have produced a partially filled tectonic depression that may have developed structural relief in excess of 7 km (Ben-Avraham et al., 1979a). The Dead Sea rift zone is considered to have a cumulative horizontal movement of 107 km and may have initially started its motion in the Senonian (Late Cretaceous) (Bartov, 1980). An interesting aspect of the Dead Sea transform fault is its very minor transverse extension (Mechie and El-Esa, 1988); however, to the east in the areas in Jordan and Syrian extrusion of alkali basalts is concurrent with faulting, and geophysical soundings indicate that the crust has been thinned by 5 to 8 km in the area underlying the volcanics.

Red Sea Trough

The Red Sea occupies a narrow elongate depression more than 2000 km long and varying in width from 180 to 300 km. The bathymetric chart of the Red Sea reveals a very distinct broad main trough (5 to 30 km wide) (Laughton, 1970). The axial trough has steep-sided fault bounded walls that form a very irregular bottom topography (Backer et al., 1975). Parallel faults mark the axial trough from Jabal Tair northward to latitude 25°N. Northward of this point the axial trough is a broad U-shaped depression lacking prominent fault scarps.

Side scan sonar of the axial trough has given a much clearer picture of the fault patterns in the inner trough (Pautot, 1983; Garfunkel et al., 1987) (Fig. 5.3). The southern part of the Red Sea reveals many subparallel faults similar to mid-ocean patterns. Spreading centers were found to be straight zones 30 to 50 km long that stepped laterally in relay zones 3 to 10 km across, without the usual transform faults shown for the Red Sea trough (Garfunkel et al., 1987). In the northern Red Sea the complex bathymetry, noted before, and crustal structure are manifested by irregular faulting that deviates widely from the axial trough trends. These faults sometimes produced a zigzag structure similar to that found on the flanks of the Gulf of Suez, suggesting control by the remaining extended older crust. The lack of obvious transform faults is puzzling and the change in obliquity does not appear to be controlled by faults. Where ocean crust formed in the northern Red Sea area it developed in a disorganized manner with many jumps between spreading centers (Martinez and Cochran, 1988). Reflection surveys reveal undisturbed sediments between the isolated spreading centers (Ross and Schlee, 1973). It is obvious from recent studies that the Red Sea has a diachronous history of spreading. The bottom structures, magnetic patterns, and reflection profiles in the different sectors point to increasing and well-established spreading centers in the southern Red Sea and, in the northern Red Sea, disorganized spreading centers with isolated "cells" forming in response to control by oblique continental stretching, with the continental blocks exerting

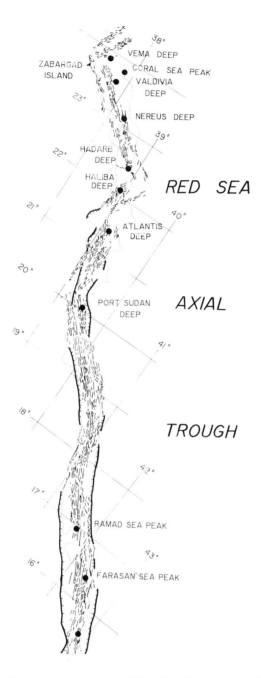

FIG. 5.3. Red Sea axial trough structures established by side scan sonar (Gloria). From Garfunkel et al. (1987), with modifications.

the predominate control (Mart and Hall, 1984; Guennoc et al., 1988, 1990; Martinez and Cochran, 1988; Egloff et al., 1991; Makris and Rihm, 1991).

Afar Depression

The Afar (Danakil) Depression represents a spectacular triple junction where the Gulf of Aden oceanic spreading center enters the Gulf of Tadjura, to meet with the north-trending East African rift system. The Red Sea trend terminates in this triangle (CNR/CNS Afar Team, 1973) (Fig. 5.4). The Afar is surrounded by high standing plateau. On the west the structural elements trend northerly for a distance of nearly 600 km. The scarp is the result of both normal and reverse faulting,

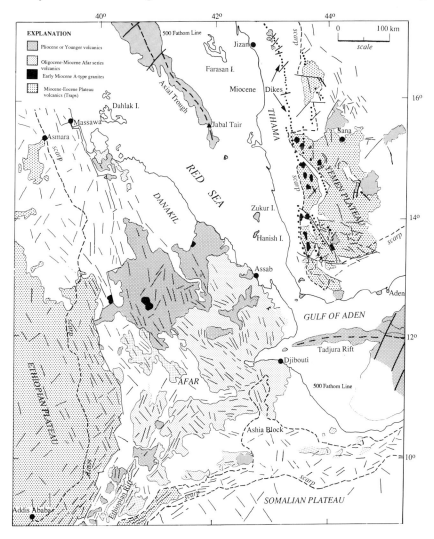

FIG. 5.4. Structural outline map of the Afar–southern Red Sea–Yemen system.

which forms a zone 25 to 50 km wide. Transverse faulting has offset the major fault zones and marginal grabens have developed with widths up to 10 km (Mohr, 1978; Kazmin et al., 1980). The southern margin of the Afar opens east-ward, changing from normal faulting at the neck to a wide zone of antithetic faults resulting from a broad zone of crustal extension (Morton and Black, 1975). The Ethiopian plateau and Afar Depression developed synchronously, in many stages, starting about 14 to 15 Ma. The Afar Depression is an emerged part of the Red Sea–Gulf of Aden structures, and is considered to be an on-land manifestation of sea-floor spreading (Barberi et al., 1972; Tazieff et al., 1972; Barberi and Varet, 1975) consisting of a thick pile of Pliocene–Pleistocene basalts superposed onto interlayered zones of older volcanics and sediments. Widespread fracturing and faulting resulting from still active extension forms belts that generally parallel the triple junction rifts that enter the depression (Backer et al., 1973; Mohr, 1978; Beyth, 1991; Courtillot, 1980; Clin, 1991). The axial zone of the Danakil is marked by active volcanoes whose chemistry is intermediate to ocean basalts and, in some cases, these basalts have fractionated to silica rich eruptives (Barberi et al., 1974). Mohr (1978) has pointed out that there are probably elements of transcurrent movement within the Afar Depression, indicating a compressive stress direction of N 30°E which, he believes, may result from ongoing separation of Arabia from Africa.

Yemen

The coastal plain within Yemen is mainly an actively alluviating surface approximately 50 km wide that covers most of the earlier Tertiary structural trends (Fig. 5.4). No major subparallel normal faults parallel to the Red Sea trend have been found by mapping even though some researchers in the past have presumed their presence (Geukens, 1966). The escarpment, marking the eastern boundary of the coastal plain, rises to more than 3000 m and is cut by oblique faults that seem to bear no relationship to the structural trends of the Red Sea. The headward erosion of the Yemen scarp is very irregular, with no obvious structural control, yet there is evidence that the recent volcanics are controlled by an east–west tectonic system (Civetta et al., 1979; Chiesa et al., 1983). The December 1982 earthquake on the Yemen plateau produced north-trending extensional cracks that followed a trend parallel to the Red Sea axis (Plafker et al., 1987). This recent activity is related to deep transcurrent faults developed within the crust and perhaps related to volcanic magmatic centers. The long volcanic history of the Yemen has covered most of the older sedimentary units, so it is now not possible to ascertain what style of faulting accompanied the volcanism during the formation of the Red Sea.

Jizan Coastal Plain (Tihama)

The coastal plain in the vicinity of Jizan consists of a series of coalescing alluvial fans derived from streams actively eroding the escarpment some 65 km inland (Brown et al., 1989). This coastal plain is essentially a featureless alluviated surface that has been undergoing active submergence since Miocene time (Gillmann, 1968). A narrow hinge zone (4 to 6 km) striking northwest parallel to the Red Sea contains the Tihama Asir igneous complex (see Fig 3.12). This zone consists of older Precambrian metamorphic rocks overlain by Phanerozoic sediments intruded by dikes, layered gabbros, and extrusive volcanics that in certain discontinuous zones are made up entirely of new igneous crust. Unconformably overlying the Precambrian basement rocks is the Wajid sandstone of Ordovician–Cambrian age, forming a series of antithetically block faulted ridges that dip towards the Red Sea. Just south of the Tihama Asir complex, near the Yemen border, Jurassic marine sedimentary rocks unconformably overlie the Wajid sandstone and have also been deformed by the antithetic block faulting. The igneous complex is dominated by a sequence of subparallel dikes that have a northwesterly strike parallel to the axial trough of the Red Sea and dip eastward towards the escarpment. The dikes are infrequent within the bordering Precambrian basement rocks exposed northeast of the hinge zone, and increase in abundance southwestward towards the axis of the hinge zone. In the hinge zone dikes have screens of Wajid sandstone, and mafic and silica volcanics, as well as the lacustrine beds of the Baid formation of the Miocene age. In restricted areas, the dike swarms have no screens of pre-existing country rock and develop chilled margins against each other. Schmidt et al. (1983) have shown that the screens made up of Baid sediments are interlayered with bimodal volcanics and contain Miocene vertebrate fossils. These sediments and volcanics are now collectively called the Jizan group (Schmidt et al., 1983).

A moderate sized funnel-shaped gabbro intrusive formed within the dike swarm, and its layers dips steeply towards the Red Sea. The layering produces a semicircular pattern at the surface, and downward projecting layering suggests a funnel-shaped body. Funnel-shaped gabbros from east Greenland are nearly identical to the Jabal Tirf gabbro and it is now generally agreed that initial dips up to 30° are formed by primary magmatic processes. The steep dips of the Jabal Tirf gabbro are probably enhanced by primary, inward dipping, igneous layering characteristic of the East Greenland mafic intrusions such as Skaergaard (Myers, 1980; Brooks and Nielsen, 1982). Fifty kilometers north of Jabal Tirf another funnel-shaped layered gabbro (Masliyah) is associated with the dikes of the hinge zone (McGuire and Coleman, 1986) (see Fig. 3.15).

These dikes and layered gabbros range in age from 18 to 26 Ma and mark the beginning of rifting and extension in this area. The Jabal Tirf igneous complex and associated rocks are distorted by extensional faulting, which was generally north–south initially and shifted westward. These extensional features are cut by dextral transfer faults that form discrete blocks (Voggenreiter et al., 1988a).

Along the central Saudi Arabian coastal plain a similar dike complex and volcanic rock association is found just south of Jiddah. The early stages of crustal extension were accompanied by faulting, graben formation, and contemporaneous dike-swarm intrusion. Subsequently, thick gabbro dikes formed. These are considered to mark the onset of rifting (Pallister, 1987), and are found all along the Saudi Arabian coastal plain from the Sinai to Yemen, representing the first significant pulse of tholeiitic magma intrusion related to Red Sea rifting.

Evidence for Extension

Interpretation of the structural development of the Red Sea Basin has focused on lithosphere stretching models (Bohannon, 1986a; Bayer et al., 1988; Voggenreiter et al., 1988a,b; Voggenreiter and Hotzl, 1989; Perry and Schamel, 1990; Bohannon and Eittreim, 1991; Cochran et al., 1991). Wernicke (1985) introduced the concept that continental stretching could be related to detachment zones that passed through the lithosphere. This mechanism could produce asymmetric configurations since half of the extending terrane is pulled from under the other. Wernicke, in his seminal paper on uniform-sense normal simple-shear of the continental lithosphere, suggested that the Red Sea rift provided geological evidence to support this idea. He related the uplift along the western margin of the Arabian plate to the upper plate dome developed above detachment faults, and described the narrow Red Sea hills on the Egyptian side as marking the edge of a steep breakaway fault. Mantle and deep crustal rocks of the Zabargad Island are considered by Wernicke (1985) to be a "core complex" exposed by low-angle normal faulting that brought these deep-seated rocks in contact with upper crustal rocks. The exposure of the Zabargad Island complex occurred because overlying crust and mantle rocks were faulted away to the east along the detachment zone. Although Wernicke was not able to carry out field studies in the region, there are now many disconnected studies that partially support the idea of continental extension as an important process in the formation of the Red Sea (Nicolas, 1985; Bohannon, 1986a; Bayer et al., 1988; Voggenreiter et al., 1988a,b; Voggenreiter and Hotzl, 1989; Bohannon and Eittreim, 1991; Mitchell et al., 1992) and even along the southern margin of Yemen and its extended margin towards the Gulf of Aden (Tard et al., 1991; Bott et al., 1992).

Evidence for normal faulting associated with continental extension was first recorded in the Red Sea area by Black et al. (1972) along the margins of the Afar Depression (Fig. 5.5). At the edge of the Somalia plateau near Dire Dawa a sequence of Mesozoic sandstones and limestones are overlain by volcanics. The sedimentary and volcanic sequence rests unconformably on Precambrian basement. This sequence is cut by a series of synthetic normal faults striking east to east-northeast, producing an area of tilted blocks whose bedding dips southward at angles between 10° and 30°. Similar normal faults with variable strikes are present on both sides of the Aisha horst some 100 km to the north. Along the western margin of the Afar, where it abuts against the Ethiopian plateau, normal faults have a northerly strike. The fault blocks are east dipping (antithetic), with their downthrown sides facing away from the Afar Depression and their bedding dipping 25° to 40° east (Black et al., 1972). The zone of faulted blocks nearly 50 km wide is covered by younger volcanics within the Afar, and is separated from the Ethiopian plateau by parallel discontinuous grabens on the west (Fig. 5.5).

Contemporaneous with this extensional faulting are sporadic occurrences of peralkaline volcanics and hypabyssal intrusives within the faulted sequences. The overlying Stratoid Series (0.35 to 11.1 Ma) rests unconformably on the block faulted sequence and fixes the upper age limit of extensional faulting (Black et al.,

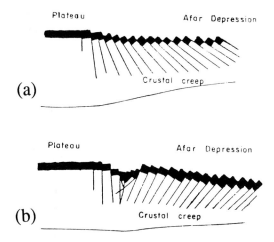

FIG. 5.5. Earliest attempt to show crustal attenuation in the Afar area. (a) Synthetic faulting and tilting of crust (Dire Dawa). (b) Antithetic faulted margin along the eastern edge of the Ethiopian plateau. From Morton and Black (1975).

1972). Black et al. also describe normal faulting in the Danakil Alps and Aisha horst, with older basalts tilted up to 45° under the cover of the younger Stratoid Series. Their conclusions led them to suggest that crustal extension had taken place, with normal faulting at the surface accommodated by ductile deformation in the lower crust. However, they were reluctant to evoke a low-angle master detachment fault beneath the normal fault system, but indicated that all of the Afar including the Danakil Alps and the Aisha horst comprised attenuated continental crust prior to the eruption of the Stratoid Series (about 11 Ma). Significant surface manifestation of normal faulting produced by extension is best recognized where block faulting cuts stratigraphically coherent sedimentary sequences. Block faulting within the complex igneous and metamorphic basement rocks typical of much of the Red Sea Basin could easily go undetected. Nearly all of the shoreline of the Red Sea consists of a complexly deformed igneous and metamorphic Precambrian basement, or is covered by Quaternary sediments. Mohr (1987) has proposed a model of rifting in the Afar region that embodies both downwarping in the initial stages and large-scale underplating of the stretched continental crust.

The discovery of synsedimentary deformation related to tectonic activity indicates that this activity may be the main factor in controlling depositional changes within the Red Sea Basin. (Purser and Hotzl, 1988; Purser et al., 1990b). These deformational structures include breccias, folds, and small synsedimentary faults and dikes, which occur in intraformational horizons where it can be shown that the deformation occurred near the sediment–water boundary (Purser and Hotzl, 1988; Purser et al., 1990b). Purser and colleagues have concluded that these features are a direct result of periodic seismicity. Their abundant occurrence at major lithologic boundaries, such as the contact between the initial rift continental sediments and the overlying deep marine clastics, as well as at the upper part of the evaporite sequence and the marine Pliocene sediments, marks major lithologic changes that could be the result of major tectonic adjustment. These structures have been correlated over distances as great as 400 km. Future stratigraphic studies will no doubt show this synsedimentary seismic deformation to be widespread and related to seismic events produced during extensional faulting (Purser et al., 1990b).

Detailed studies of the Tihama coastal plain near Jizan have provided additional evidence for continental extension (Coleman et al., 1979; Bohannon, 1986a; Voggenreiter et al., 1988a,b; Bayer et al., 1989; Bohannon and Eittreim, 1991, (see Figs. 3.12, 3.15, and 5.6). Along the narrow coastal strip, Paleozoic Wajid sandstone rests unconformably on the metamorphic and igneous Precambrian basement and is overlain by Mesozoic sandstone and limestone. These rocks are intruded by mafic dikes, gabbros, granophyres (Jabal Tirf igneous complex) and are in part overlain by the Jizan group sediments and volcanics (Schmidt et al., 1983). These units within the area generally strike parallel to the Red Sea trend (north to northwest) and dip 20° to 60° towards the Red Sea. Wide-spaced normal faults, east of the Tihama coastal plain within the scarp area, offset Paleozoic sandstones with displacements around 100 m (Coleman, 1984b; Bohannon, 1986a). These faults dip 60° to 80° to the east and form blocks down to the Red Sea, representing the most eastern surface expression of continental thinning. Within the coastal plain a series of faulted Paleozoic sandstone ridges, trending northwest-southeast, dip 50° southwest, is cut by synthetic normal faults dipping 20° to 40°. These faults repeat the sandstone section but the lack of stratigraphic markers in the massive sandstone precludes an accurate estimate of the offset (amount of extension) (see Figs. 2.2, and 2.3). These synthetic normal faults are truncated by low angle faults situated along the contact between the sandstone and the Precambrian basement. Bohannon (1986a) maps such faults as detachment faults related to surficial sliding of the sandstone blocks. Steep normal faults that repeat the section are thought to flatten downward towards a hypothetical east dipping major basal detachment fault extending to mid-crustal depths (Bohannon, 1986a). A similar scenario is postulated by Voggenreiter (1988a,b); however, he considers that the detachment fault is a major zone extending into the mantle below the western edge of the Arabian margin, following Wernicke's (1985) original suggestion. Transgression of the Tertiary sediments across the coastal plain obscures the nature of the extensional complex towards the Red Sea, making it difficult to establish the limits of extension into the Red Sea Basin.

These studies of the Tihama coastal plain provide evidence for early continental extension in this part of the Arabian continental margin. Whether extended continental crust exists beyond the Tihama coastal area cannot be settled at this time. The age of the extension (normal faulting) is closely related to the generation of the Tihama Asir complex, which ranges in age from 22 to 25 Ma. Normal faulting has clearly offset and tilted dikes, as well as the large gabbro body. However, later mafic dikes invade and cross-cut the normal faulting, suggesting that faulting and magma invasion were almost contemporaneous. There is abundant structural evidence that the early stages of rifting are marked by both normal faulting and invasion of mafic magma.

The Gulf of Suez can be used as a model for the early development of the Red Sea Basin, as its development was arrested by the Dead Sea fault in the Early

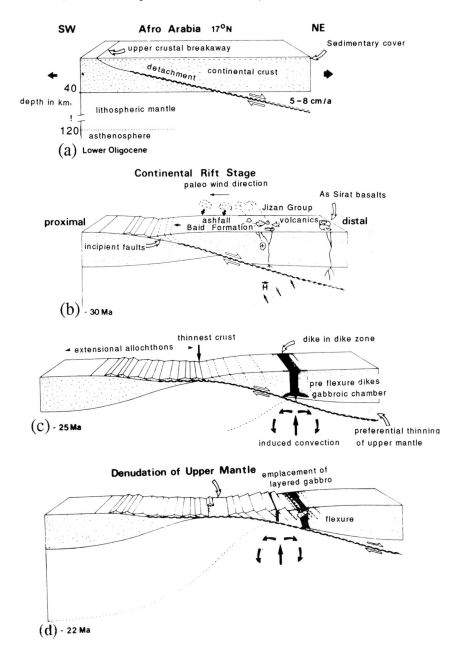

Fɪɢ. 5.6. Hypothetical early stages in normal simple shear model for the evolution of the southern Red Sea. From Voggenreiter et al. (1988b), following the initial idea of Wernicke (1985).

Miocene (Garfunkel et al., 1987) (Fig. 5.2). Extensive oil exploration data, including reflection and refraction seismic data and basin stratigraphy from numerous exploratory wells, provides a wealth of information (Ott d'Estevou et al., 1987; Beydoun, 1988; Evans, 1988; Said, 1990) (Fig. 5.7). The Suez basin appears to consist of two offsets and superimposed basins, perhaps as a result of the major faults following curvilinear patterns (Bosworth, 1985). These basins are asymmetric, and show evidence of rapid subsidence. Normal faulting parallel to the basin axis produced block faulting and minor rotation within the separate half graben basins. Within the individual half graben the dip is in the same direction, either northeast or southwest parallel to the Suez basin axis. Within the basin

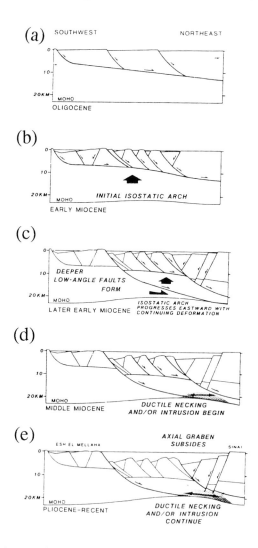

Fig. 5.7. Proposed model for extension in the southwestern part of the Gulf of Suez. From Perry and Schamel (1990).

there are three distinct dip provinces, changing polarity across complex transfer zones. Longitudinal changes to the southeast are found in increasing dips of the pre-rift sediments, fault-bounded blocks decrease in size, and calculated extension increases from 5 to 20 km. These variations are related to greater counterclockwise rotation of the Sinai Plate at its southern end (Colletta et al., 1988).

The development of basaltic volcanism and subparallel dikes in the Suez basin area 25 to 23 Ma is almost identical to the period of rifting and igneous activity at Jabal Tirf. The period of rapid extension (about 19 Ma) in the Gulf of Suez was confined to the narrow axial zone, which was the only area unaffected by later uplift. After the Suez rift was decoupled from the Red Sea by the Dead Sea fault, subsidence decreased and extension virtually ceased (Evans, 1988). Uplift and erosion of the rift flanks continues due to a deep seated thermal perturbation produced by mantle melting and underplating of the crust (Steckler, 1981; Buck, 1986; Steckler et al., 1988). The failed Suez rift will slowly stabilize as the thermal perturbation declines, and it is unlikely that oceanic spreading will be established (Steckler, 1985; Buck et al., 1988).

Further evidence of continental extension around the margins of the Red Sea requires additional geophysical data and exploratory drilling. However, the evidence available at this time supports the idea of continental extension in the early stages of Red Sea rifting. Brittle extension by faulting, either by simple shear or by pure shear, was accompanied by significant amounts of contemporaneous magma underplating and invasion of the thinning crust. All models of the development of the Red Sea will have to recognize the contemporaneity of normal faulting and invasion of magma.

Zabargad Island

This small island is located south of Ras Banas on the Egyptian side of the Red Sea, approximately 50 km west of the axial trough. Its structural importance relates to the fact that it represents an uplifted fragment of lower crust and upper mantle (Bonatti et al., 1981, 1983; Styles and Gerdes, 1983; Nicolas et al., 1987; Bonatti, 1988) (Fig. 5.8). Regional structural relations indicate that it may be part of a major tectonic zone (the Zabargad Fracture Zone striking N15°E) parallel to the Aqaba fault, located at the point where the strike of the Red Sea axial trough rotates approximately 40° to the east (Bonatti et al., 1984). Numerous normal faults subparallel to the Zabargad Fracture Zone may be related to extension of the Egyptian continental margin.

Peridotite, consisting of fresh lherzolite and amphibole-bearing lherzolites, is the main exposed rock type. These peridotites are considered to have formed initially within the mantle, but were altered and hydrated by tectonic uplift and deformation related to the Miocene rifting (Bonatti et al., 1986; Piccardo et al., 1988). Amphibolites interlayered with felsic gneisses in tectonic contact with the peridotites also show similar overprinting due to the Miocene rifting event (Boudier et al., 1988; Seyler and Bonatti, 1988). Isotopic studies using $^{143}Nd/^{144}Nd$ and $^{86}Sr/^{87}Sr$ indicate that these rocks have a Sr–Nd isochron age of about 675 Ma and a Pan-African mantle signature (Oberli et al., 1987; Brueckner et al., 1988). Basalt dikes cut the older basement rocks and are mafic intrusions formed in the Miocene stage of rifting (Petrini et al., 1988). Foliations in the peridotite are northwest–southwest, and vertical lineations produced by stretching plunge 50° northwest (Nicolas et al., 1987). The preserved feldspathic lherzolites exhibit a dextral high T shear flow that Nicolas et al. (1987) consider to be a result of mantle diapir upwelling. A low temperature sinistral shear flow produces mylonitic textures in the amphibole-bearing amphibolites and is thought to represent emplacement of these rocks during the Miocene rifting event (Nicolas et al., 1987) (Fig. 5.8). K/Ar ages of about 23 Ma on the amphiboles (Villa, 1988) in the peridotites and amphibolites support the idea that this fragment of Pan-African lower crust and mantle was emplaced during the Miocene rifting event (Nicolas et al., 1987).

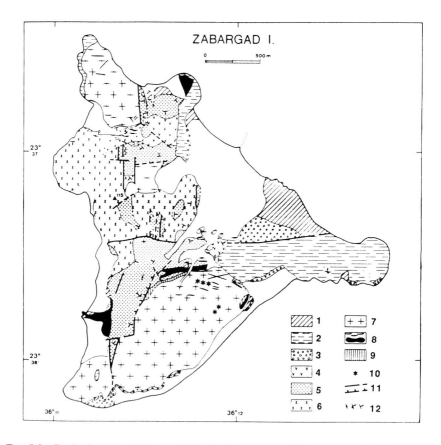

Fig. 5.8. Geologic map of Zabargad Island. (1), young reef limestones; (2), old reef limestones; (3), breccia and conglomerates; (4), evaporites; (5), Zabargad sedimentary formation; (6), metamorphic complex; (7), peridotites; (8), basaltic–doleritic dikes and shallow intrusions (Miocene); (9), gabbroic rocks; (10), nickel mineralizations; (11), faults; (12), dips of foliation. From Bonatti (1988).

As shown above, the proposed model by Voggenreiter et al. (1988a,b) following the simple shear extension ideas of Wernicke (1985) and Lister et al. (1991) permits stretching and thinning of the crust combined with concomitant passive upwelling of the upper mantle. Early Miocene extensions are manifested all along the Red Sea Basin, mainly by mafic dikes and listric faulting (Bohannon, 1986a; Voggenreiter et al., 1988b; Bohannon and Eittreim, 1991). The single outcrop of the mantle–lower crust at Zabargad Island provides evidence that both continental extension and mantle upwelling are combined with invasion of magmas into the lower crust.

6

Geophysical Outline

The Red Sea Basin has received much attention from the geophysical community because it represents a newly formed ocean basin developed as part of the present day movement of the Arabian plate away from Africa. Most of the data has been gathered by shipboard techniques because it is much easier to obtain permission for oceanographic studies than for onshore investigations. Major subcrustal structures cut across territorial boundaries, but few traverses of these boundaries have been possible in nearly all of the studies carried out. It is hoped that future studies will benefit from a more cooperative political climate.

This section is written from a geological viewpoint, so some of the more important nuances near and dear to the geophysicist's heart may be missing! What is most important is that geophysical studies have provided a wealth of new data, which strongly modulate *ad hoc* geologic ideas.

Magnetics

The utilization of measurements of the Earth's magnetic field to understand the subcrustal nature of the Red Sea crust has produced significant results (Phillips, 1970; Girdler and Hall, 1972; Girdler and Styles, 1976; Miller et al., 1985). Hall et al. (1977) summarized earlier magnetic field data by producing a residual magnetic anomaly map of the Red Sea, incorporating magnetic measurements from aerial and sea surveys. This map provides basic data that has been utilized by numerous scientists trying to elucidate models concerning the formation of the Red Sea. A detailed survey of the southern Red Sea (19° to 15°N latitude) was published at around this time (Roeser, 1975), but its data were not incorporated into the Hall et al. (1977) map.

The most striking area of the magnetic anomaly map is centered within the axial trough of the Red Sea, where large amplitude (up to 800 nT), sharp peaked anomalies (usually less than 15 km wide) produce linear patterns that parallel the northwest trend of the Sea. This coherent linear pattern seems to terminate at 23.5°N latitude, and at 15°N latitude near Jabal Tier in the south (Hall et al., 1977).

Within the main trough away from the central axial zone of the Red Sea much smaller amplitude (300 nT) anomalies are found, with very broad wavelengths up to 30 km wide. These anomalies are found irregularly on both sides of the axial trough and may extend over the flanking coastlines. In the northern Red Sea isolated anomalies form a pattern apparently unrelated to the axial trend (Hall et al., 1977).

Areas encompassing the continental crustal areas have variable amplitude anomalies (50 to 400 nT) with restricted asymmetric wave lengths (5 to 30 km). These anomalies are more characteristic of continental shield areas and are quite distinct from the axial trough and main trough magnetic anomalies.

A system of prominent north–northwest trending narrow negative and positive areomagnetic anomalies with steep gradients are present in a narrow zone (about 100 km wide) from Jizan to the Gulf of Aqaba. These long narrow anomalies are related to nearly vertical dikes (5 to 50 m wide) of tholeiitic affinities (Blank, 1977).

The linear anomalies within the Red Sea axial trough have been interpreted as being developed from basaltic rocks formed in an oceanic spreading center (Girdler and Styles, 1974; Roeser, 1975; Hall et al., 1977). Roeser's detailed magnetic study surprisingly revealed asymmetric spreading in the southern part of the Red Sea, with the African side some 70% faster than the Arabian side which produces a change in spreading direction more nearly parallel to the axial trough (1975). In the northern Red Sea, a spreading rate of 0.55 cm per year is predicted on the basis of the reconstruction of McKenzie et al. (1970) but, as Roeser (1975) points out, spreading rates lower than 0.55 to 0.60 cm per year cannot sustain a true magmatic oceanic spreading center, but may be more typical of continental extension and magmatic intrusion.

There is disagreement as to the nature of the axial trough spreading center magnetic anomalies when considering the low amplitude broad anomalies within the main Red Sea trough. Girdler and Styles (1974) and Hall et al. (1977) point out that up to 4 to 5 km of sediments overlie the basement of the main trough, thus reducing the magnetic intensity of the underlying basement magnetic anomalies. Utilizing single magnetic profiles across the southern Red Sea, a comparison of the broad low amplitude magnetic anomalies with synthetic profiles was made. Synthetic profiles from 25.43 to 26.86 Ma and from 35.00 to 37.61 Ma provide enough similarity to enable utilization of either time period. On the basis of these comparisons Hall et al. (1977) and Girdler and Styles (1974) suggest that two distinct stages of ocean crust formation characterize the Red Sea basement (Fig. 6.1). One period of spreading is dated 41 to 34 Ma, and the other is from about 4 to 5 Ma to the present. Frazier (1970) and Cochran (1983) provide evidence that at least some of the low-amplitude magnetic anomalies can be correlated with basement faulting related to extension of the offshore area near the Dahlak Islands. The localized and linear nature of the low-amplitude magnetic anomalies in the southern Red Sea does not continue to the northern part of the Red Sea main trough.

Cochran (1983) has pointed out that the smooth low-amplitude circular magnetic anomalies in the northern Red Sea are also coincident with positive gravity anomalies. He suggests that these are mafic intrusions, either as sills up to 3 km thick invading evaporites or as deeper cylindrical-shaped intrusions invading underlying Precambrian crust. As pointed out by Coleman (1984b), these intrusions probably develop during extension and may underplate the extending crust as well as invade it. The linear magnetic anomalies shown in Gettings (1977) along the Tihama Asir coastal plain near Jizan are related to the partially exposed mafic intrusive complexes, and represent new crust invading and underplating the extended Precambrian crust in the early stages of the development of the Red Sea Basin. The Tihama Asir coastal plain aeromagnetic lineaments are collinear with or parallel to the large single dikes extending into the Sinai, and represent a major invasion of tholeiitic magma. The age of these dikes (22 Ma) constrains the age of the initial rifting to Early Miocene and would favor using the 25.43 to

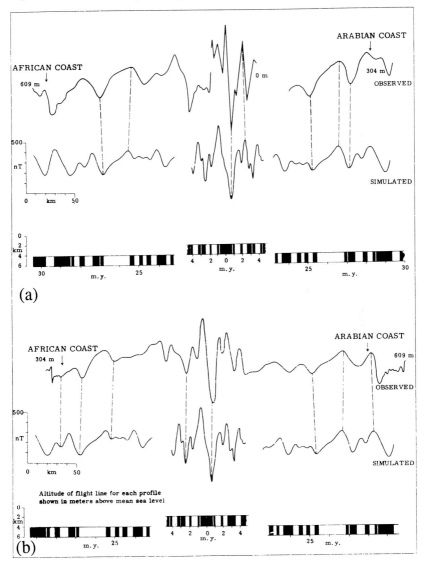

FIG. 6.1. Comparison of observed and simulated magnetic anomaly profiles across the Red Sea. From Hall et al. (1977). (a) Latitude 16°N, (b) Latitude 19°N.

26.86 Ma synthetic magnetic profile of Girdler and Styles (1974) for the broad amplitude magnetic trends within the main trough of the Red Sea as part of the first stage of Red Sea rifting.

Gravity

There have been many gravity measurements within the Red Sea (Vening Meinesz, 1934; Girdler and Harrison, 1957; Allan et al., 1964; Drake and Girdler, 1964; Worzel, 1965; Allan and Pisani, 1966; Allan, 1970; Gouin, 1970; Makris et al., 1970a,b; Quershi, 1971; Plaumann, 1975; Gettings, 1977; Hall et al., 1977; Yousif, 1982; Gettings et al., 1986; Cochran et al., 1991). A recent Bouguer gravity map compiled by Makris et al. (1991a) for the Red Sea has combined much of the data into an important database (Fig. 6.2). As with the magnetic map of the Red Sea, the most prominent gravity anomalies are present within the axial trough, form-ing elongate discontinuous anomalies of 100 to 150 mGal. It is suggested that these anomalies are caused by the isostatic effects of the ocean crust (Andy Griscom, personal communication). Even though these anomalies are discontinuous the central axial zone gravity high is continuous and never falls below 40 mGal. Like the magnetic anomalies, the gravity anomalies follow the narrow and deepest part of the Red Sea, where active sea-floor spreading has been continuous for the last 4 to 5 Ma. By matching the free-air anomalies with the gravity effect of the axial zone topography and calculating different density contrasts, an average value of 2.83 g cm^{-3} was estimated for the axial zone igneous complex (Plaumann, 1975). This estimated density is nearly the same as the value found for MORB basalts 2.86 g cm^{-3}, further reinforcing the concept that the axial trough consists almost entirely of newly formed oceanic crust..

In contrast to the axial gravity high along the main trough, a series of subdued gravity lows and highs paralleling the axial trough are present on both sides (Yousif, 1982). The presence of a thick sedimentary section (up to 5 km), increas-ing in thickness away from the axial trough and consisting mainly of evaporites, can explain these gravity lows of −80 to −110 mGal. Variable topographic relief may also contribute to these anomalies. As pointed out in the previous section, these sediments attenuate any magnetic signature of the basement, and the gravity anomalies are similarly attenuated by the blanket of sediments. The actual nature of the basement rocks beneath this cover therefore remains controversial. Explorat-ory drilling for oil in these thick sedimentary sections has not yet penetrated deep enough to resolve this problem.

A detailed gravity survey across the Saudi Arabian coastal margin in the Jizan area (Gettings, 1977) is the only area where there is a tie between the sedimentary basin of the Red Sea and the continental margin. In the trough area just offshore from Jizan negative Bouguer values of −30 mGal over the Farasan Islands extend eastward across the Tihama Asir coastal plain (Fig. 6.3). At the inner edge of the coastal plain, a discontinuous row of elongated gravity highs (+25 to +35 mGal) are caused by partially exposed layered gabbro and mafic dike swarms of the Jabal Tirf complex (Gettings, 1977). The Early Miocene age (about 22 Ma) of

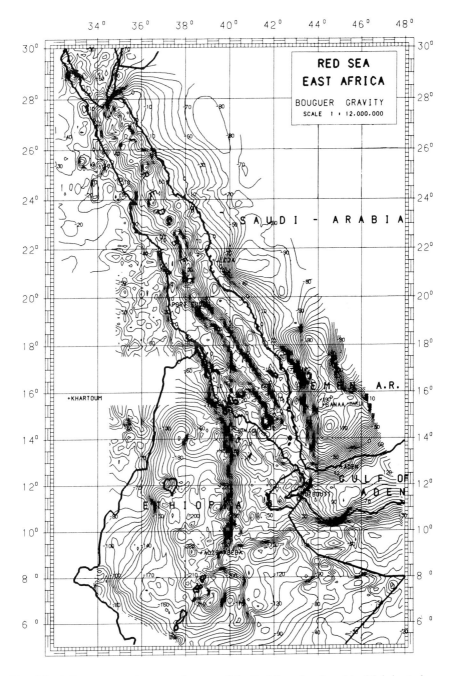

FIG. 6.2. A Bouguer gravity anomaly map of the Red Sea, developed by Makris et al. (1991a).

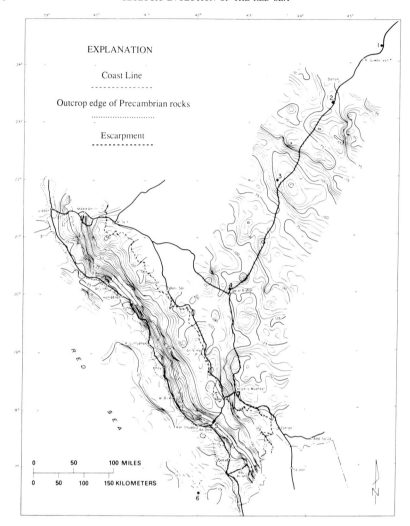

FIG. 6.3. Bouguer gravity map of the U.S. geological survey geophysical transect across the Arabian margin and the Tihama coastal plain. Contour interval is 5 mGal. Note the steep gradient marking the edge of the continent and the high anomalies along the coast from Jizan northward, marking probable gabbro intrusions related to underplating. From Gettings et al. (1986).

these intrusions ties them to the large single dikes that extend northward to the Sinai and southward into the Yemen (Blank, 1977). At the eastern boundary of these intrusions and their associated gravity anomalies, an extremely steep negative gravity gradient (4 to 5 mGal km^{-1}) marks the transition from the Red Sea Basin to Precambrian continental crust to the east. This boundary is marked by outcrops of the continental crust with an average Bouguer anomaly of –100 mGal (Gettings,

1977; Gettings et al., 1986, and Girdler, 1991). Girdler and Styles (1974) consider that this boundary marks the edge of Red Sea mafic oceanic crust, and conclude from this configuration that nearly all of the Red Sea is underlain by mafic oceanic crust, from shoreline to shoreline (Gass, 1977). Modeling of this configuration by Gettings (1977) requires a triangular wedge of basaltic material 20 km thick (2.85 km cm^{-3}) at the continental boundary and thinning westward to normal oceanic crustal thickness (8 km) below the Farasan Islands. A more attractive scheme to account for this mafic wedge at the continental margin is to call on underplating and invasion of basaltic magma into the extending continental crust (Coleman and McGuire, 1988).

Comparison of the gravity data from the southern Red Sea area emphasizes that caution should be exercised in proposing a model for all of the Red Sea based on data and observations from only one area. Cochran et al. (1991) present data on the section of the northern Red Sea from 26.30°N latitude to 28.30°N latitude where free-air gravity anomalies form subparallel highs and lows parallel to the northwest trend of the axial trough (these anomalies may be artifacts of the continental margin edge effects). These elongate anomalies range from 50 to 70 km long, and their axes are parallel to the axial rift and often located near the fault scarps within the axial zone. It is important to point out that in the axial trough of the northern Red Sea area Miocene evaporites and post-Miocene pelagics are continuous across the trough. The well-defined lineated magnetic anomalies characteristic of the southern Red Sea are lacking (Cochran et al., 1991). The Red Sea type deeps in the northern part are discontinuous and are often associated with normal and reversed magnetized dipolar magnetic anomalies, coinciding with positive gravity anomalies characteristic of shallow mafic plutons at these latitudes (Hall et al., 1977). Using gravity data combined with seismic reflection lines and heat flow measurements, a model of the crustal structure beneath the Red Sea has been developed along a single profile at 27°N latitude (Cochran et al., 1991). A synthetic, symmetric crustal profile was constructed at this latitude and utilized for a series of calculated gravity profiles. The calculated gravity anomalies were found to be 25 mGal more negative within the axis than the actual measured gravity data. This difference could be adjusted to fit the observed gravity profile by introducing contemporaneous igneous underplating or by magmas intruding the extended crust within the axial zone (Cochran et al., 1991). Modeling data in this fashion has given alternate modes of stretching the transitional nature of the crust in the northern Red Sea. Age constraints are not available for the introduction of mafic magmas in the development of the transitional crust but it is quite conceivable that underplating may be a dominant feature of the early magmas, which were too heavy to rise higher into the overlying, extended crust. Eventually, as extension continued, some magmas rose higher into the less dense extended crust, forming shallow intrusions with dipolar magnetic and positive gravity signatures that are found today in the northern Red Sea.

Comparison of gravity data with Red Sea observations from the Afar Depression is interesting. The Afar Depression represents attenuation of continental crust in a transitional state similar to the Red Sea, but in its present configuration it is exposed above sea level. The Bouguer gravity map of northern Ethiopia displays

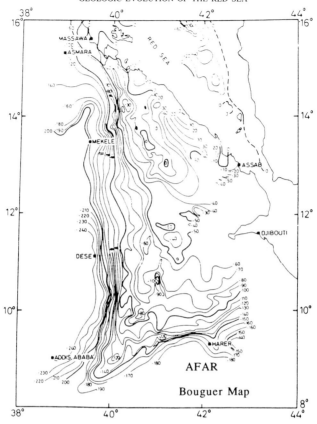

FIG. 6.4. Gravity map of the northern Afar. Note the steep gravity gradient on the western and southern boundaries of the Afar with the surrounding plateau. From Makris et al. (1972).

a close adherence to the geomorphology of the region (Makris et al., 1975) (Fig. 6.4). The high Ethiopian lava plateau have their gravity minimums situated at the highest elevations, up to –250 mGal in Dese decreasing to –185 mGal in the southwest plateau area which is comparable to the Saudi Arabian scarp minimum of –180 mGal. These trends follow the isostatic effects of topography, indicating a thick continental crust under the plateau, which thins considerably over the Afar Depression. The gravity gradient between the Ethiopian plateau and the adjacent Afar Depression varies from 2 to 3 mGal km^{-1}, decreasing to 1 to 1.5 mGal km^{-1} down the southern escarpment to the depression, and is not quite as steep as the 4 to 5 mGal km^{-1} gradient reported by Gettings (1977) in the Tihama area of Saudi Arabia.

The gravity increases with lower elevations, reaching zero gravity at the Red Sea coastline near Massawa. The Danakil mountains on the northeastern side of the Afar rise to 1700 m and coincide with a –60 mGal anomaly very similar to the Aisha horst anomaly (–65 mGal) on the southeastern flank of the Afar Depression, reflecting the presence of thicker continental crust in these specific areas. The northern part of the Afar is dominated by northwest trending anomalies parallel to the Red Sea and centered over the central volcanic ranges, but these diminish to –35 mGal over the salt plains of Dallol, considered to be an evaporite basin about 3km thick (Hutchinson and Engels, 1970; Tierclelin et al., 1980). In the southern part of the Afar the gravity anomalies bend south–southeast and can be traced to the Gulf of Tadjura (Ruegg, 1975; Jobert, 1980). In the extreme south, the Wonje fault zone is marked by a series of maxima extending up to the central Afar and marking the termination of the Ethiopian rift. Within the Afar Depression no high amplitude gravity zones comparable to the Red Sea axial rift (+100 to +150 mGal) or the Tihama Asir intrusives (+25 to +30 mGal) are present, supporting the concept that oceanic crust has not yet been created in the Afar. Modeling of the Afar crust using seismic refraction data (Berckhemer et al., 1975) shows the Afar to be continental crust attenuated to about 20 km, but in the more active extensional zones thickness is reduced to 14 km. Under the Ethiopian scarp the crustal thickness reaches 42 km, diminishing westward to 30 km in the Sudan. The transitional crust of the Afar Depression and the northern Red Sea have many features in common (Beyth, 1991). This analogy serves to show how complex the basement of the Red Sea may have been prior to syn-rift sedimentation.

Seismic Reflection

Numerous studies have been made in various parts of the Red Sea using shipboard reflection surveys, so the shallow structures are quite well known (Knott et al., 1966; Phillips and Ross, 1970; Ross and Schlee, 1973, 1977; Searle and Ross, 1975; Uchupi and Ross, 1986; Martinez and Cochran, 1988; Richter et al., 1991; Rihm et al., 1991) (see Fig. 2.9). These results define a consistent relationship for the central part of the Red Sea. The central axial zone is well defined but devoid of sediments, and shows a rough topography characteristic of ocean crust formed at spreading centers. Underlying the main trough and the shelf a widespread, unique subbottom acoustic reflector (S) is present some 0.2 to 0.7 seconds below the seafloor (see Fig. 2.9). Drilling has shown that the sediments overlying the S-reflector are Late Miocene to Late Pleistocene in age and consist of clayey silt and ooze (Ross et al., 1973). The S-reflector consists of a thick section of evaporites (up to 5 km) and is the same age as the M-reflector in the Mediterranean, which also marks the end of the salinity crisis (Hsu and Bernoulli, 1978). Normal faults are abundant in the main trough area with the downthrown sides facing the axial trough. Some reverse faults are present and minor folding related to salt diapirism is common. Deformation of sediments below the S-reflector is much more intense than that seen in the sediments above. The normal faults, which are considered to be the result of continuous extension across the Red Sea Basin, propagate upward and displace the S-reflector surface in many areas. Invasion of the evaporites and overlying pelagics by infrequent volcanism away from the axial trough indicates that not all volcanism is related to the spreading center of the axial trough.

In the southern part of the Red Sea (south of 15°N latitude) and in the northern part (north of 26.30°N latitude) the S-reflector is continuous across the axial trough (Guennoc et al., 1988; Cochran et al., 1991) (see Fig. 2.9). No reflections characteristic of oceanic crust were detected, even though there is normal faulting of the sedimentary sequence. Even though the sedimentation rate is higher for the post-Miocene sediments in the southern Red Sea the S-reflector can still be seen clearly identified in seismic reflection profiles that cross the axial zone.

The acoustic basement of the axial trough can be traced outward towards the shoreline for only 60 km. Within the main trough most of the reflection surveys have not been able to penetrate the thick evaporite cover. However, within the inner parts of the trough the evaporite section thins and reflection surveys in the mid-part of the Red Sea reveal a continuous reflector that rises rapidly towards the coastline (Yousif, 1982).

Reflection profiles provide evidence for the shallow structures of the Red Sea and support the magnetic and gravity evidence that the ocean crust occupies the central axial trough (about 120 km), but it is not possible to image the basement beyond the axial trough. The basement reflectors under the shelf indicate a sea-ward dipping uniform slab that could be interpreted either as stretched continental crust or as oceanic crust. Drilling will be required to distinguish between the two choices.

Seismic Refraction

As described above, the acoustic shipboard surveys of the shallow crust of the Red Sea have been successful in detailing structures and unconformities in the upper 1 to 2 km of the basin, because at such shallow depths it is easy to produce enough energy by air-gun techniques. Controlled explosion seismic experiments are capable of estimating velocity, structure, and composition of the crust and upper mantle to depths of 100 km. However, the structural detail obtained by reflection techniques cannot be matched because of the difficulty in distinguishing reflection from refraction energy arrivals at recording points, and because fewer receivers and greater distances are involved (Mooney and Prodehl, 1984). Some of the refraction seismic lines accomplished in the Red Sea area are shown in Fig. 6.5 (Davies and Tramontini, 1969, 1970; Tramontini and Davies, 1969; Makris et al., 1970a, 1975, 1981; Ginzburg et al., 1979a,b; Makris et al., 1970a, 1983b; Prodehl, 1985; Avedik et al., 1988; Bayer et al., 1989; Prodehl and Mechie, 1991). These studies have provided reasonable estimates of continental crustal thick-ness and margins around the basin. Because of the rise of the asthenosphere within the rift, elevated temperatures within the crust and mantle may produce unpredictable variations in seismic velocities. The danger of confusing hot mantle with high velocity lower crust is a problem, as is the unpredictable correspondence of seismic velocity with different rock types. Furthermore, the seismic ray tracing often covers distances of several hundred kilometers, whereas rocks over those distances may be made up of small-scale mixtures of disparate rock types, yielding seismic velocities unrelated to the chosen rock.

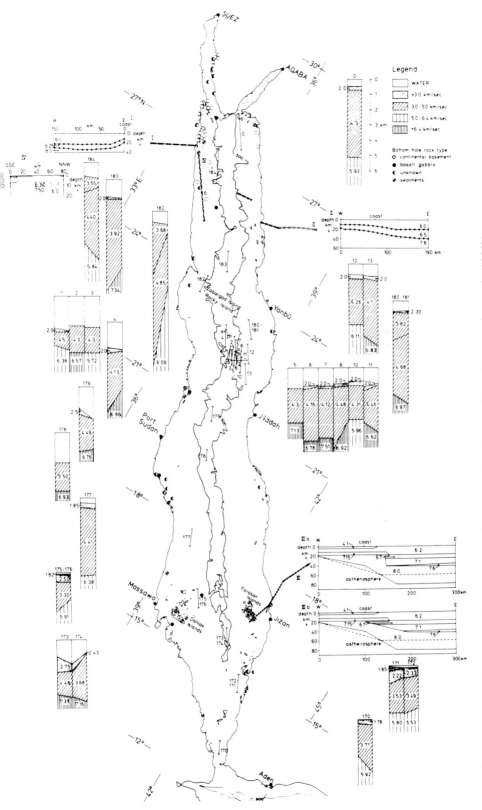

FIG. 6.5. Summary of refraction profiles accomplished within the Red Sea area. From Voggenreiter et al. (1988b).

The seismic reflection line recorded by the U.S. Geological survey team in Saudi Arabia provides data from six shot points along a 1070-km-long northeast-trending line that originates within the main Red Sea trough at the Farasan Islands and extends across the continental margin at Jizan and to the vicinity of Riyadh (Healy et al., 1982; Mooney and Prodehl, 1984; Gettings et al., 1986). The importance of this line is that it crosses the transition between the Red Sea Basin and the edge of the Saudi Arabian continental crust. Furthermore, the data generated by the experiment were shared with seismologists as part of the IASPEI Commission on controlled source seismology, to produce several independent interpretations (Mooney et al., 1985) (Fig. 6.6). This was very enlightening for geologists, who learned that great variations in interpretations could evolve, depending on the technique of refining the raw data.

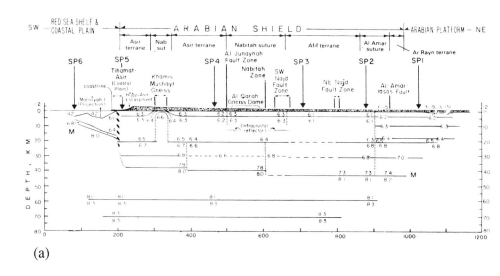

(a)

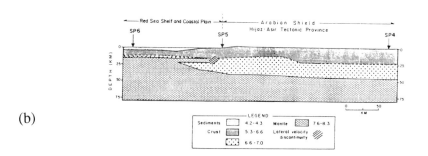

(b)

FIG. 6.6. (a) Suggested crustal section for the Arabian margin from seismic refraction profile. From Gettings et al. (1986). (b) Alternate crustal section using the same data as (a). From Milkereit and Fluh (1985).

The seismologists participating in these comparisons generally agreed on the crystalline crust of the Arabian Shield being about 37 to 44 km thick, with the upper crust having velocities of 5.85 to 6.45 km s^{-1}. A discontinuity between 10 and 20 km appears to separate the upper crust from the lower crust, and the Moho is marked by upper mantle velocities of about 8.0 km s^{-1}. The lower boundary may be too thick, as Pleistocene xenoliths indicate anomalously high heat flow in the upper mantle. This high heat flow could reduce the upper mantle velocities enough to give erroneous depths for the Moho (McGuire and Bohannon, 1989).

A much more divergent set of interpretations results from the modeling of the transition zone (Fig. 6.6a,b). According to Gettings et al. (1986) the Moho discontinuity decreases from 38 km under the shield to 20 km at the western shield margin (Fig. 6.6). From the edge of the shield to the Red Sea margin a simple three-layer crust is proposed, with Tertiary sediments forming a basin up to 4 km thick but thinning in both directions away from the basin axis. The velocity of the sediment is considered to be 4.1 km s^{-1}, including both the clastics and evaporites. Underlying the sedimentary basin is crustal material which thins from 17.5 km at the coast to 8 km under the Farasan Islands. This wedge-shaped body is given velocities from 6.0 to 6.4 km s^{-1} and contains a 2 km base whose velocity is 6.8 km s^{-1}. Gettings et al. (1986) suggest that the average velocity for this wedge is 6.2 km s^{-1}, which is intermediate between oceanic crust layers two and three, and well within the ranges given for diabase.

An alternate interpretation by Milkereit and Fluh (1985) suggests that the margin of the Red Sea has a shallow crust–mantle boundary at about 15 km and is overlain by 3 to 4 km of sediments, thickening to 6 km along the axis of the basin (Fig. 6.6b). The underlying basement to the sediments is uniform and is considered to be either oceanic crust with a thickened layer two or modified continental crust extended and invaded by mafic magmas. At the edge of the shield an inversion is proposed where the Moho remains at a constant depth of 15 km, but a second Moho at 20 km is present under the deepest part of the sedimentary basin, reaching a depth of 40 km just under the shield margin. According to Milkereit and Baarakt (1985), this uniform mantle layer (about 7 km thick) can explain the strange Pn and PmP (seismic signals) arrivals, as well as the steep Bouguer gravity recorded at the very edge of the shield. Mechie and Prodehl (1988) discuss these two opposing models but provide no insight as to which of them might be preferred.

A petrological assessment of both models is needed to evaluate the interpretations. Even though the velocities and gravity provide a good fit, the oceanic crust is much thicker at the margin than any ordinary oceanic crust, so the Gettings et al. (1986) model diverges from known oceanic crustal thicknesses. The Mechie and Prodehl (1988) model does not explain the double Moho in terms of structure, where flat-lying compressional thrusting is required in a tectonic setting considered to be extensional. An alternate suggestion would involve early rifting at the shield margin (about 30 Ma), where MORB-like magmas are developed by shallow mantle melting but cannot migrate through the lighter upper crust and would either intrude or underplate the extending continental crust. When these magmas reach a neutral buoyancy at various levels they would produce a denser and thicker crust, particularly along the continental margin boundary as revealed by the positive Bouguer anomalies associated with a crust whose velocities are similar to diabase and gabbro. Mid-ocean style ocean crust spreading cannot be established until the spreading rate is fast enough to allow the mafic magmas to reach the ocean-bottom–seawater interface. The Red Sea Basin sediments with evaporite at densities of 2.00 g cm^{-3} and pelagics at 2.30 g cm^{-3} would act as density barriers for the rising basaltic liquids, whose density is 2.63 g cm^{-3}. Eroded exposures of dikes and gabbros along the Red Sea coastal plain reveal shallow level intrusions that reached neutral buoyancy high in the continental crust. The long-term argument as to oceanic crust versus new mafic magma intruding extending crust cannot be solved by geophysical methods alone. The source and composition of the magma is the same in both cases and is not controlled by the style of formation, i.e., whether developing at a seawater interface, invading overlying sediments, or extended crystalline basement.

The Afar region has been covered by a series of five seismic lines varying from 120 to 250 km long, with a single line 300 km long over the Ethiopian plateau (Berckhemer et al., 1975). The Ethiopian plateau line reveals a crustal thickness of 40 km near the crest of the eastern scarp, diminishing to 35 km westward in Sudan. Here the upper mantle velocity is 8.0 km s^{-1}, with the lower crust velocity of 6.6 to 6.7 km s^{-1} at a depth of 15 km, overlain by upper crust with a velocity 6.2 km s^{-1}, similar to the velocity structure of the Saudi Arabian continental crust. Within the Afar Depression the upper layers, consisting of volcanics and sediments, have a thickness of up to 6 km and velocities from 2.8 to 4.1 km s^{-1}. The basement rocks of the Afar are not true oceanic or continental crust (Gouin, 1970) and can best be explained as a mixture of extended continental crust invaded by ponded mafic magma (Mohr, 1989). The upper layer is 2.7 km thick with velocities of 6.0 to 6.1 km s^{-1}, underlain by a lower crust 10 to 20 km thick with velocities of 6.6 to 6.8 km s^{-1}. These measurements are quite similar to those found by Gettings et al. (1986) along the continental edge in Saudi Arabia. In the Afar Depression the Moho has been determined to be at a depth of 18 to 26 km, with the upper mantle showing velocities of 7.3 to 7.7 km s^{-1}. With these reduced upper mantle velocities it is clear that there is active asthenosphere rising beneath the Afar, either by convection or as a rising plume.

The Jordan–Dead Sea rift has had two separate seismic soundings carried out normal to the rift (Ginzburg et al., 1979a,b; El-Isa et al., 1986, 1987a,b). These results demonstrate that a strike–slip mechanism concentrated in the upper 21 km of the brittle crust is the main source of earthquake energy release within the Dead Sea rift. A regional uplift of 1 to 2 km is associated with crustal thinning down to 33 to 36 km, where the underlying mantle has anomalous velocities of 7.95 to 8.10 km s^{-1}. Besides the dominant strike–slip motion there is an asymmetric extrusion of Late Tertiary to Holocene basalts on the eastern flank, which requires partial melting in the mantle brought about by crustal thinning and rise of the asthenosphere under the rift.

Numerous other refraction lines have been carried out (Makris, 1983a,b); Makris and Rihm, 1987; Gaulier et al., 1988) in the northern Red Sea, where thickness of 40 km appears normal for both the African and Arabian crust, with crustal thinning to less than 20 km across the continental margins. Sedimentary sequences up to 4 km thick extend all across the northern Red Sea, with no thinning across the axial trough. Basement velocities from 5.8 to 6.93 km s^{-1} are similar to other values found within the main trough. These velocities have been interpreted as either oceanic crust, extended continental crust, or extended continental crust invaded by MORB-like basaltic magmas. In the southern Red Sea there are too few traverses to draw any specific relationships. Berckhemer et al., (1975) east–west model of Afar demonstrates that there is crustal thinning near Assab towards the Red Sea. The one offshore refraction profile (Girdler, 1970b) just east of Assab shows more than 2 km of sediments overlying extended continental crust with an estimated velocity of 5.92 km s^{-1}. Thus there is no hard geophysical evidence indicating the presence of oceanic crust in the axial zone of the southern Red Sea. Rather, it would appear that continental extension is distributed over the 400 km from the Ethiopian scarp to the Yemen plateau scarp.

Seismic Tomography of the Red Sea Region

Improvements in the analyses of seismograms has provided new insights into the three-dimensional structure of the mantle in the Red Sea region (Romanowicz, 1991). Tomography is a method by which images obtained from worldwide networks of digital seismographs are inverted to reveal inhomogeneties within the mantle. These variations in velocity can then be used to produce three-dimensional maps of the mantle seismic velocity anomalies that can be related to tectonic features (Aki et al., 1977; Tanimoto and Anderson, 1985; Anderson, 1990; Romanowicz, 1991). Because the sampling of seismograms is uneven, due to the erratic distribution of earthquakes, numerous methods of parameterization have been utilized to develop workable tomographic models (Romanowicz, 1991). Modeling techniques so far developed have revealed important correlations between surface crustal tectonic features, such as faster velocities under cratons and slow velocities under oceanic spreading centers and hot spots.

Zhang and Tanimoto (1992) have produced a new global S-wave velocity model utilizing Love and Rayleigh waves from 18,000 seismograms, which has greatly improved the resolution of seismic tomography. Previously, horizontal resolution of anomalies was in the order of 4000 to 6000 km but Zhang and Tanimoto can now distinguish anomaly horizontal lengths of 800 to 1000 km and vertical depth lengths from 70 to 80 km near the Earth's surface, diminishing to 300 km at depths of 500 km. Utilizing these results from a tectonic viewpoint, the variation of S-wave velocities results from variations in temperature (mantle convection) and presence of a liquid (partial melting).

Two depth slices crossing at right angles above the Afar hot spot, constructed by Zhang and Tanimoto (1992), provide the first images of the mantle underlying the Red Sea Basin. These depth slices include data extending 500 km into the mantle and more than 3000 km away from the hot spot. This data produced image information on the mantle under the surrounding cratons as well as on the full

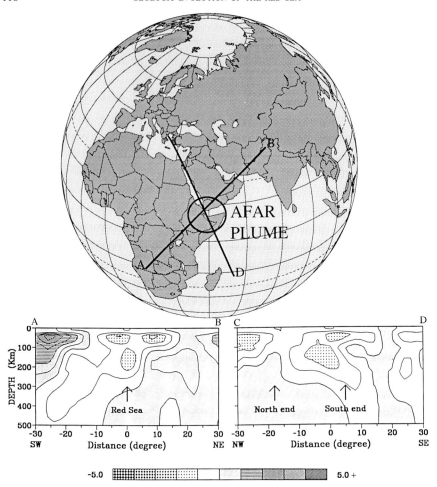

F<small>IG</small>. 6.7. Depth slices of S-wave velocity parallel to (CD) and normal to (AB) the Red Sea. The bold circle outlines the Afar hot spot, with a diameter of 1000 km. The location of the slices is shown on the globe view, with the distance from the crossing to the end points about 30° (3330 km). Velocity anomalies are given in steps of 1% from 0 to 5 (+ or –) with contours within each step of 0.5 %. Note that the crossing of the depth slices is at 14°S latitude and 42°E longitude not centered on the Afar hot spot. The azimuth for AB is 50° and for CD 151°. From Zhang and Tanimoto (1992).

length of the Red Sea (Fig. 6.7). These depth slices display the velocity anomalies, given as a percentage difference (+ or –) of the global average at each depth. They are emphasized by hatched areas, representing percentages, and are internally contoured in 0.5% intervals.

Interpretation of these depth slices has just begun and will undoubtedly become more sophisticated as more data is accumulated. Zhang and Tanimoto (1992) present depth slices for spreading ridges, showing symmetrical, shallow-low-velocity anomalies extending less than 100 km into the upper mantle. In comparison, the depth section parallel to the Red Sea (CD) reveals an irregular pattern of low-velocity extending to depths greater than 200 km, similar to hot spot patterns but with a pronounced flattening southward under the Horn of Africa. The extension of the low-velocity anomaly obliquely to depths greater than 300 km indicates a connection with the Afar hot spot (Zhang and Tanimoto, 1992). The African east–west depth section (AB) at right angles to the Red Sea trend reveals an asymmetric low-velocity anomaly extending below 200 km obliquely under the eastern side of the African craton. To the east an isolated low-velocity anomaly under the Arabian craton indicates magma generation at depths of 100 km or more, which is supported by the xenolith studies on the Arabian harrat lavas (McGuire, 1988b). These equivocal results indicate that a hot spot exists in the Afar at present, but it is not clear whether the passive spreading now taking place in the central part of the Red Sea can be connected with the Afar anomaly. The resolution of the Zhang–Tanimoto model, even though a great improvement, is not enough to distinguish clearly the thermal effect of passive Red Sea spreading from the hot spot influence. These results do, however, reveal the fact that these low-velocity anomalies extend under the Arabian and African plates, perhaps demonstrating erosion and melting of older continental lithosphere by invasion of melts upwelling from either hot spot activity or new passive spreading centers, initiating magmatic underplating. As pointed out by Zhang and Tanimoto (1992), these low-velocity anomalies require a much broader upwelling than a small diameter plume (100 to 300 km) (Schilling et al., 1992) and therefore may be an expression of multiple coalescing plumes or a combination of passive mantle upwelling associated with plate motion enhanced by the Afar hot spot. Earlier chemical studies preclude a single source for the erupted magmas of the Red Sea area. Continued accumulation of seismic data from this area will improve the new and important depth slices presented by Zhang and Tanimoto.

Heat Flow

Measurement of heat flow provides important data on the recent history of thermal processes within the Red Sea Basin. An early comprehensive review by Girdler (1970b) was based on only 12 measurements from oceanographic vessels and deep oil exploration wells. These early measurements clearly showed that heat flow is a function of distance away from the axis of the Red Sea (Girdler, 1970b; Girdler and Evans, 1977). Thus it was shown that the Red Sea pattern is consistent with other world rift systems and is probably related to the underlying mantle convection and consequent volcanic eruptions in the axial trough and along the flanks of the Red Sea. In 1976, Scheuch produced the first heat flow map of the Red Sea using all of the older data and new data from *R.S. Valdiva* cruises. He was also one of the first to show that variable heat flow values in the axial zone indicate discontinuous spreading along the axis. Makris et al. (1991c) have produced a new heat flow density map of the Red Sea based on a total of 467

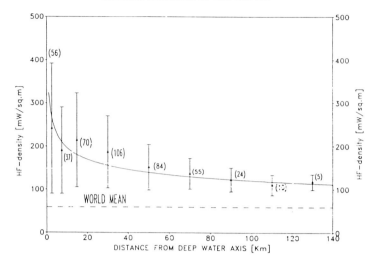

Fig. 6.8. Averaged heat flow curve developed from measurements of the Red Sea. These values are averaged over perpendicular distance from the axial trough without respect to the north–south orientation. From Makris et al. (1991c).

heat flow values (Fig. 6.8). In addition, new measurements are available from the Dead Sea rift (Ben-Avraham et al., 1978; Ben-Avraham and Von Herzen, 1987) and the Gulf of Aden (Haenel, 1972). There has been a remarkable increase in the amount of data, but it is still inadequate to develop properly an overall thermal picture because of the concentration of data points in the axial zone and in the northern Red Sea.

A number of technical problems are involved in the acquisition of data, but the Makris group has utilized all available data to produce the map. These average heat flow values are plotted in Fig. 6.8, showing that the flanks of the Red Sea trough have values of 116 mW m^{-2}, well above the world average of 59 mW m^{-2}. The axial zone has extreme variations of heat flow because of differences in elevation and convective systems, but shows average maximum values of 400 mW m^{-2} and an average of 200 mW m^{-2} for the whole axial trough (Makris et al., 1991c). The symmetry implied by this average plotted curve of all the Red Sea heat flow measurements is not attained when single detailed profiles are plotted separately (Verzhbitskiy, 1980; Gettings et al., 1986; Martinez and Cochran, 1989; Makris et al., 1991c). Interestingly, the few heat flow measurements of the Precambrian crust away from the Red Sea margins are usually lower than the world average. Morgan et al. (1980) report values of about 40 mW m^{-2} for the western desert of Egypt and an average of about 74 mW m^{-2} for nine measurements in the eastern desert. Heat flow measurements carried out during a seismic refraction traverse across western Arabia reveal that heat flow values are well below the world's average (59 mW m^{-2}) some 300 to 900 km east of the axial trough (Gettings et al., 1986).

All of these measurements were obtained at the surface and represent the present state of thermal equilibrium, but may not reflect present day thermal conditions in the mantle. McGuire and Bohannon (1989) compared the surface heat flow of the Arabian shield to the values estimated for the mantle directly under the heat flow measurements profile of Gettings et al. (1986). Using mantle xenoliths from the youngest eruptive AOB, (0 to 5 Ma), McGuire estimated temperatures of 900 to 1000° C at depths of 40 to 50 km. Assuming present day surface heat flow values and that a geotherm based on 40 mWm^{-2} extended to the same depths where the xenoliths equilibrated produces a large discrepancy of +450° C, which led McGuire to conclude that partial melting is probably taking place at about 80 km in the mantle where a much higher geotherm (90 mWm^{-2}) intersects the peridotite solidus (McGuire and Bohannon, 1989). Calculations of the potential surface heat flow as a function of lithosphere thinning to depths of 80 km requires that the actual thinning is probably younger than 20 Ma, otherwise surface heat flow anomalies would appear over the western Arabian shield (McGuire and Bohannon, 1989). This discrepancy between the surface and the mantle heat flows gives new insights into the relationship between mantle convection and the present day axial trough, which is supported by the preliminary tomography map of the mantle in the Red Sea area (Zhang and Tanimoto, 1992). It is possible to suggest that there is an apparent asymmetry of the mantle convection under the African and Arabian plates at their margins.

A detailed study of three heat flow profiles in the northern Red Sea (Martinez and Cochran, 1989) produced several computer models that test the relationship between a simple shear model and several pure shear models, utilizing a two-dimensional time-dependent numerical technique that follows the advection and diffusion of heat. The models produced the following constraints on the development of the northern Red Sea: (1) simple shear extension above a shallow dipping detachment could not match the observed heat flow and could not produce thermal anomalies high enough to allow shallow mantle melting; and (2) a pure shear extension also falls short of producing heat flow values high enough to cause melting. Horizontal gradients may be present at the edges of the rift, producing convection, which might generate additional heating similar to that discovered by McGuire and Bohannon (1989) and supported by the tomography.

A calculation of the isotherms using surface heat flow measurements and superimposed on a synthetic seismic refraction profile from the Suakin deep to the African margin produces very shallow high temperatures (850°C) in the axial trough that gradually descend to only 20 km under the thinned crust of the African margin (Makris et al., 1991c). These abnormally high values at the thinned base of the lithosphere along the African margin are remarkably similar to the values derived by different parameters for the Saudi Arabian shield (McGuire and Bohannon, 1989).

Makris et al. (1991c) warn that "models based only on heat transport by conduction, neglecting the convection in the mantle and water circulation in the sediments, can provide only very rough estimates for the distribution of the temperature with depth and time." The use of geophysical measurements in models is a powerful tool for testing geological ideas but such models are only two-dimensional instantaneous "windows" into complicated three-dimensional situations with time variant parameters.

Geophysical Constraints

1. Magnetic anomalies within the central axial zone of the Red Sea provide a clear record of sea-floor spreading in this area for approximately the last 5 Ma. Broad amplitude magnetic anomalies away from the axial zone cannot be successfully interpreted as representing earlier formed ocean crust.

2. The newly compiled gravity map of the Red Sea basin indicates dense oceanic crust in the central axial zone and in the Gulf of Aden. Steep gravity anomalies along the margins of the Red Sea and the Afar Depression indicate rapid changes in crustal thickness. Such thickening of the crust may be the result of magmatic underplating. Crustal extension and magma injection produce ambiguous gravity signatures in the Afar Depression and on the margins of the Red Sea.

3. Seismic reflection lines and side scan sonar reveal active normal faulting in the axial trough. The S-reflector present in all parts of the Red Sea basin marks the beginning of sea-floor spreading and the end of the salinity crisis.

4. Seismic refraction lines across the margins of the Red Sea indicate crustal thinning in a narrow zone. Interpretations of the thinning mechanisms are diverse, and there is no consensus.

5. Heat flow measurements give a picture of increasing surface heat flow from the margins of the Red Sea to the axial trough, supporting the concept of present day ocean spreading in this area. Differences in estimated heat flow values for the surface of the Saudi Arabian craton and its underlying mantle indicate active mantle convection.

7

Red Sea Plate Tectonics

The present movement of Arabia away from Africa has fascinated many earth scientists and has provided a large and varied selection of papers dealing with the plate tectonics, using various kinematic indicators. This chapter has nothing new to offer in improvement over the various models already presented. I have attempted to integrate all of this information into a summary of our knowledge.

Kinematics

Integration of the Red Sea neotectonics into a global plate tectonic system has met with moderate success because a kinematic reconstruction of plate movement requires that plates remain rigid and that motion on each boundary is somehow related to the motion on all others (Le Pichon et al., 1973).

At the outset of plate tectonics, numerous kinematic models were proposed for the plate motions within the Red Sea area. The synthesis of McKenzie et al. (1970) clearly and accurately provided an acceptable model, which has subsequently been modified without significant improvement (Girdler and Darracott, 1972; Le Pichon and Francheteau, 1978; Gass, 1980; Berhe, 1986; Courtillot et al., 1987; Hempton, 1987; Joffe and Garfunkel, 1987). These particular kinetic models are based on the instantaneous motion of the plates and pertain mainly to the very recent history of the Red Sea.

It is generally accepted that the area can be divided into four rigid plates: **Arabia, Africa (Nubia), Sinai,** and **Somalia**. These plates are bounded by the Red Sea, the Gulf of Aden, the Dead Sea fault (transform), the East Africa–Afar rift, and the Suez rift. Intersections of these boundaries in the Afar triangle and at the northern end of the Red Sea form triple junctions (McKenzie et al., 1970; Le Pichon and Francheteau, 1978; Joffe and Garfunkel, 1987) (Fig. 7.1).

Even though the plate boundaries may be diffuse such as at the Gulf of Suez and in the Afar triangle, assumptions are made that certain points along plate boundaries can be used to define motion between contiguous plates. In order to estimate motion directions on individual plates Eulerian poles of rotation can be established by shoreline matching or by movement along faults. The total motion model of Joffe and Garfunkel (1987) is perhaps the best yet proposed for the plate motions to account for the last 4 to 5 Ma, however, reconstruction of plate motion older than 5 Ma in the Red Sea area has not been successful.

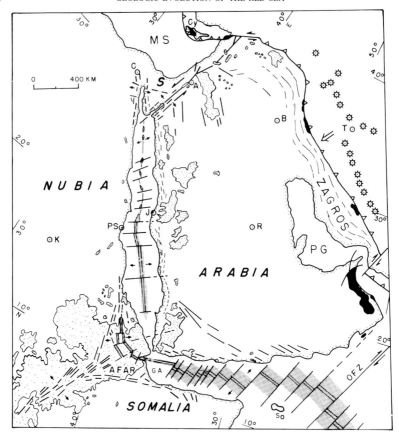

FIG. 7.1. Plate tectonic setting for the Red Sea showing the main structural features and the distribution of Cenozoic volcanic rocks. The double lines show presently active spreading centers, with double arrows indicating spreading direction. The estimated areas of new oceanic crust formed by spreading are shown in heavy stipple. Volcanic fields are represented by light stipple. The asymmetry of volcanism can be ascertained easily on this map. Major faults are shown in heavy lines, those with barbs indicate thrusting; the direction of movement is shown by arrows. Crustal attenuation is also shown by opposite arrows in the Afar and the northern gulfs of the Red Sea. Stars indicate subduction related calc-alkaline volcanics during the late Tertiary consumption of Arabia under the Eurasian plate east of the Zagros fold belt. Short dashed lines along the Red Sea coast and in the Sinai indicate dike swarm trends. A, Amman; B, Baghdad; C, Cairo; Cy-Cyprus; GA, Gulf of Aden; J, Jiddah; K, Khartoum; MS, Mediterranean Sea; OFZ, Owen fracture zone; PG, Persian Gulf; PS, Port Sudan; R, Riyadh; S, Sinai; So, Socotra; T, Tehran.

Assuming the total Red Sea opening was about a pole situated at 32.5°N, 24.5°E the motion of its northern end would have to be 125 to 140 km. A movement of 105 km along the Dead Sea transform is about a pole located near 32.7°N, 19.8°E (Garfunkel, 1981) and accounts for the post–Miocene movement and extension of the Suez rift of 30 to 40 km. The pole for the Arabian–Somalian plate separation is centered on 26.5°N, 21.5°E and is also generally accepted as the pole of the Gulf of Aden opening (Cochran, 1981; Joffe and Garfunkel, 1987). The motion in the Afar triangle is controlled by a diffuse triple junction where separation of the African–Somalia plates is poorly constrained, and the plate separation here is less than 15 km (Courtillot, 1980). The widespread crustal heating and hot spot activity has combined to produce diffuse crustal stretching over a zone 400 km wide extending from the Ethiopian scarp to the Yemen plateau.

It is possible to produce a reasonable kinematic reconstruction of the Red Sea area using rigid plate tectonic assumptions for the last 4 to 5 Ma. Prior to this time the various plate boundaries were not rigid because of extension and thinning. The Red Sea Basin, as has been shown earlier, cannot consist of oceanic crust from shoreline to shoreline (Cochran, 1983; Coleman, 1984b; Voggenreiter et al., 1988a). The accumulated evidence shows extentional normal faulting is accompanied by mafic intrusion and underplating demonstrating that the Red Sea basement has been modified prior to the onset of sea-floor spreading. The Gulf of Suez provides a good model for the early extension within the rift zone, where normal faulting and restricted mafic intrusions mark the earliest stages of rifting. The Red Sea can be divided into three separate sections whose history is distinct. In the northern Red Sea, the presence of restricted magmatic centers distributed in a random fashion within the basin indicates that continental extension and mafic intrusions are contemporaneous (Cochran and Martinez, 1988; Guennoc et al., 1988). The central Red Sea shows continuous sea-floor spreading since 4 to 5 Ma, but thick sedimentary cover obscures the pre-Pliocene history of the basement (Cochran, 1983). In the southern Red Sea, it is clear that considerable extension has taken place without production of oceanic crust. This extension reaches its maximum in the area from the Yemen plateau to the Ethiopian plateau, including the Afar triangle. Numerous attempts have been made to reconstruct microplates within the Afar but there is no evidence that such plates existed for any length of time (CNR/CNS Afar Team, 1973; Barberi et al., 1974a; Mohr, 1978; Beyth, 1991). The Danakil Alps, an exposed portion of extended continental crust on the edge of the Afar Depression, has rotated during extension and is perhaps a small block of extended continental crust that escaped large-scale magmatic intrusion during extension, and therefore remains as a more buoyant block.

The Gulf of Aden provides further evidence that the boundaries of the plates underwent a period of crustal thinning prior to steady state ocean floor spreading, which was initiated 10 Ma (Laughton, 1966; Girdler, 1970a, 1983; Girdler and Styles, 1978; Courtillot, 1980; Cochran, 1981, 1983; Tard et al., 1991). The age of the oceanic crust in the Gulf of Aden indicates that it is propagating toward the Afar triangle rather than away from it (Courtillot, 1980). A similar situation exists in the southern Red Sea, where new crust is propagating southwards toward the Afar rather than away from it, assuming that the Afar triangle represents an active plume during the last few million years (Schilling, 1973).

It is apparent from the above discussion that the plate boundaries of the Red Sea area have earlier histories of extension and magmatism that cannot be reconciled by long-term finite rigid plate movements. On the other hand, a kinematic model of plate movement can be developed for the last 4 to 5 Ma.

Early Rifting

Considerable controversy surrounds the early manifestations of rifting within the Red Sea area. The presence of a large, pre-rift, Arabian–African dome was proposed by Gass (1970). The dome is centered over the Afar Depression and its uplift is thought to have been initiated in the Oligocene or perhaps earlier. Contrary to this, numerous papers present evidence that such a dome did not exist, but instead propose that this same area was low-lying and near sea level from the Late Cretaceous to the Early Oligocene (Coleman, 1984a; Schmidt and Hadley, 1985; Bohannon et al., 1989; Brown et al., 1989). The presence of Early Tertiary–Late Cretaceous sediments is reported in a number of areas around the Red Sea (Brown et al., 1989). As described earlier in Chapter 2, a Cretaceous–Paleocene narrow embayment extended southward to Jiddah subparallel to the present Red Sea axis. The shallow-water sediments of this embayment are distal equivalents of the Tethyan ocean situated to the north (Dercourt et al., 1986). Bohannon et al. (1989) have suggested that land surfaces within the hypothetical dome were generally low-lying, with laterites forming within a highly vegetated lowland. These laterites have been preserved under the basal flows within Saudi Arabia (Coleman et al., 1977; Schmidt et al., 1983; Pallister, 1987) and the basal volcanics of the Ethiopian and Yemen plateau rest on laterite or reworked laterites. Thickness contours on the pre-rift eruptions of the Ethiopian and Yemen plateau indicate that these eruptions were extruded within a downwarping basin (Pilger and Rosler, 1976). The Hail Arch situated on the Arabian plate forms an elongate north–south structure, which is thought to have originated during the Late Cretaceous (Greenwood, 1973), although others suggest an earlier age (Powers et al., 1966). These observations preclude the existence of a large hot-spot dome forming in the early stages of the Red Sea–Afar–Gulf of Aden triple junction, as has been suggested by Schilling (1969, 1973) Gass, (1970), White et al. (1987), White and McKenzie (1989) and Schilling et al. (1992).

A general regression of the marine sedimentation during the Late Eocene and Oligocene age is related to a low sea-level stand and is not the result of up-doming (Almond, 1986a). The geologic evidence therefore shows that the earliest phases of the Red Sea Basin formation took place on a subdued low-lying surface near sea level. Northwest-trending faults over this broad area may have been the first manifestation of regional extension (Almond, 1986a). Even though there was localized doming subsequent to these early extensions, there is no obvious surface uplift that could be related to a hot spot centered in the Afar triangle (Mohr, 1982).

Syn-Rift Events

The Early to Late Oligocene eruptions of alkali basalts were concentrated within the Afar–Ethiopian–Yemen area with minor eruptions extending northward into

Saudi Arabia. The early eruptives were transitional in composition and their feeder systems were controlled by northwest fracture systems parallel to the present axis of the Red Sea. In nearly all cases the basal flows of these eruptives covered preserved laterite surfaces or shallow-water marine sediments deposited in a southern extension of the Tethys ocean (Bohannon et al., 1989). Siliceous non-marine and volcanoclastic sedimentary rocks (Baid formation) associated with mafic and silicic flows of Early Miocene age (19 to 21 Ma) containing freshwater fossils are present in the Jizan area. These rocks mark the earliest deposits within the Red Sea Basin and are a manifestation of bimodal volcanism associated with lake sediments (Schmidt et al., 1983; Brown et al., 1989). These same continental lacustrine and fluvatile sediments are present in the Dogali series of Ethiopia and the Hamamit formation in coastal Sudan (Savoyat and Balcha, 1989; Bunter and Abdel Magrid, 1989a). In the northern Red Sea similar continental sediments interbedded with volcanics and considered to be Oligocene have been described by Purser et al. (1990a) and Purser and Hotzl (1988). Inter-layered coarse clastics and volcanics in the Gulf of Suez (Nukhul formation) demonstrate that the earliest sediments within the Red Sea Basin were probably deposited in separate morphotectonic depressions (Evans, 1988). Some geologists have suggested that the early rifting in the Red Sea Basin was controlled by strike–slip faulting with the formation of pull-apart basins. Thus, the fluvatile and lacustrine sediments may have been restricted to isolated basins initially developed by strike–slip motion with the development of pull-apart Basins (Ott d'Estevou et al., 1987; Makris and Rihm, 1991). Hughes et al. (1991) have suggested an inter-mittent Oligocene connection between the Red Sea and the Indian Ocean across the Afar, and propose Oligocene rifting in the southern Red Sea which may be earlier than that observed in the Jizan area.

Contemporaneous with these pull-apart basins was a very widespread emplace-ment of tholeiitic dikes which was accomplished in an extremely short time span (Sebai et al., 1991). These dikes extend from the Sinai to the southern Tihama in Yemen (about 1700 km) and are chemically distinct from the AOB flows found to the east in Yemen and Saudi Arabia. The dike rocks are formed from mantle magmas produced by large amounts of partial melting, and in some areas make up significant areas of new crust (passive margin ophiolite) (Coleman, 1984b). These mafic intrusions have two distinct modes of occurrence: (1) tholeiitic dike swarms forming sheeted sequences, with concomitant development of cone-shaped magma chambers without significant screens of older continental crust, and (2) wide and elongate dikes of differentiated tholeiitic magma, forming extensive linear but noncoalescing dike swarms. It has been estimated that these dikes may make up to 5% new crust and if they extend downwards into the mantle, their volume could be greater than $100,000 \text{ km}^3$. Refraction sections along the Arabian coast indicate thicknesses of mafic crust up to 18 km under the Arabian margin (Gettings et al., 1986). It is conceivable that underplating of these mafic magmas contributes to the unusually thick section at the continental margin. There is evidence that normal faulting associated with extension was contem-poraneous with invasion of these magmas (Voggenreiter et al., 1988a). Intrusive and extrusive activity of a similar age is found on the African plate but the volume is so much less that the aysmmetry of the exposed volcanics is obvious (Thrope and Smith, 1974; Ressetar et al., 1981; Steen, 1982; Almond et al., 1984; Almond, 1986b).

These intrusives represent new crust forming from magmas similar to those that produce oceanic crust at spreading centers. Geophysical measurements on the basement rocks under the flanks of the Red Sea reveal magnetic, gravity, and seismic velocity signatures that are compatible with the concept that this flanking basement could be considered oceanic crust (Girdler and Styles, 1974; Gettings, 1977; Labrecque and Zitellini, 1985). It is suggested that basement rocks of the Red Sea Basin flanks are made up of a mixture of extended continental crust invaded by tholeiitic magmas, whose shape and form is variable along strike-producing geophysical signatures that can be mistaken for the more organized spreading of oceanic crust. During this period of rifting (30 to 5 Ma) alkali basalts were erupted in the plateau areas and tholeiitic magmas invaded the extended crust in the Red Sea Basin. The magmas in the Red Sea Basin would not reach the surface because the extension rate was too slow but instead would consolidate in the basin as they reached a neutral buoyancy within the enclosing sediments (Ryan, 1987). Only when the extension rate is great enough (1 cm per year $^{-1}$) will oceanic crust form at the seawater interface well below the freeboard of the continental margins. When the extension rate is too slow rising magmas would invade the extended crust and stagnate in the overlying unconsolidated sediments at a much higher level where a state of neutral buoyancy could exist (Bonatti, 1985).

During the period of rifting and extension of the continental crust, sediments from the Gulf of Suez extending into the southern Red Sea have similar stratigraphies. Lowering of global sea level in the Early Miocene isolated the Red Sea and Mediterranean, allowing only small incremental surges of fresh seawater. Sediments consisting mainly of marine deposits with increasingly thick sections of evaporites began to form (Stoffers and Ross, 1977). During this period evidence for flank uplift along the Red Sea is signaled by the presence of coarse polymictic conglomerates being deposited along the margins of the Red Sea basin (Coleman, 1984b). Fission track ages suggest that this uplift began about 13.8 Ma (Middle Miocene) and is continuing up to the present time, with a maximum uplift of 2.5 to 5 km (Kohn and Eyal, 1981; Omar et al., 1987; Bohannon et al., 1989).

The rifting within the Red Sea Basin and Gulf of Suez is accompanied by normal listric faulting, first established by reflection seismic profiles and then by geologic mapping in areas where Mesozoic sedimentary rocks are exposed (Morton and Black, 1975; Voggenreiter et al., 1988a,b; Bohannon et al., 1989; Savoyat and Balcha, 1989). Unconformities and faulting of the syn-rift sedimentary series revealed by seismic profiling and exploration drilling provide evidence that extension has been a continuous process during rifting (Lowell and Genik, 1972; Miller and Barakat, 1988; Savoyat and Balcha, 1989; Mitchell et al., 1992).

The combined geologic evidence indicates that the early syn-rifting in the Red Sea was non-uniform and involved both continental thinning and igneous intrusion, producing a continental margin strongly modified by thermal and mechanical overprints. Any attempt to reconstruct kinematic movements by rigid plate tectonics will probably be unsuccessful. The single boundary where estimates of plate movements can be made is along the Dead Sea transform extending more than 1000 km from the Gulf of Aqaba to the Tuarus–Zagros convergence zone. A total left lateral offset of 105 km has been established (Freund, 1965; Bartov, 1980; Eyal et al., 1981). Offset of the 20 Ma tholeiitic dikes in the Sinai by the Dead Sea transform limits its initial movement to the Early Miocene. As the Dead Sea transform developed, extension within the Gulf of Suez diminished, with accelerated uplift of its flanks. The constraint on the opening of the northern Red Sea is tied directly to the Dead Sea transform and precludes a coast to coast opening of the Red Sea.

Sea-Floor Spreading

The widespread unconformity within the Red Sea Basin sediments marking the shift from evaporite basin sedimentation to open ocean marine sedimentation is the most dramatic marker for the initiation of sea-floor spreading (Stoffers and Ross, 1977). Seismic reflection profiles show widespread truncation of reflectors in the marine oozes above the evaporite section which has been interpreted as a major Pliocene–Pliestocene unconformity (Ross et al., 1969). The axial trough fissure eruptions extrude on the inner floor within the deepest part of the graben, producing new oceanic crust. These areas of new oceanic crust have strong, short wavelength linear magnetic anomalies in the central and southern parts of the axial trough (Girdler and Styles, 1974; Roeser, 1975; Hall, 1989). Various inter-pretations of time reconstructions indicate that sea-floor spreading began 5 to 6 Ma with rates of 0.5 to 1 cm per year (Girdler and Styles, 1974; Roeser, 1975; Labrecque and Zitellini, 1985; Izzeldin, 1987; Hall, 1989). Other areas over the main trough show weak, long wavelength anomalies which have been correlated with an earlier period of sea-floor spreading (Girdler and Styles, 1974; Girdler, 1985; Girdler and Underwood, 1985; Hall, 1989). As suggested earlier, these magnetic anomalies represent a mixture of extended continental crust with non-uniform mafic intrusions which cannot be used to produce reliable chronological sequences comparable to uniform oceanic crust sequences. The well-defined magnetic anomalies of the Red Sea axial trough narrowing to the north and to the south outline a very restricted area of the Red Sea from about 17°N to 23.5°N which is underlain by oceanic crust. Kinematic reconstructions will have to incorporate rifting (continental extension) both in the north and the south to accommodate the Red Sea total finite opening. The initiation of sea-floor spreading in the central and southern Red Sea marks resurgence of AOB eruptions within the Arabian plate as well as increased activity in the Afar Depression. At this same period the Dead Sea transform changed its style of deformation, with a marked component of transverse separation leading to pull-apart basins and local volcanic eruptions along the transform (Joffe and Garfunkel, 1987). The northern collision of the Arabian plate with Eurasia within the Biltis suture continues (Lovelock, 1984; Lyberis et al., 1992), but with a change in direction so that eruptions on the Arabian plate follow rifts with a north–south orientation oblique to the Red Sea trend. This configuration continues up to the present, with historic eruptions of AOB on the plateau and continued submarine eruptions in the axial trough producing many separate hydrothermal systems, and actively concentrating heavy metals.

Constraints on Red Sea Plate Reconstructions

1. Rigid plate boundaries were modified by continental thinning.
2. Kinematic reconstruction beyond 4 to 5 Ma is not possible.
3. A hot spot generated dome did not exist prior to early rifting and volcanism.
4. Flank uplifts are related to magmatic underplating during early rifting.
5. Early stages of continental rifting produced isolated basins by a combination of normal and strike–slip faulting.
6. Sea-floor spreading began 4 to 5 Ma and is propagating north and south in the Red Sea. Sea-floor spreading initiated in the Gulf of Aden 10 Ma is now propagating westwards towards the Afar hot spot.

8

Economic Aspects of the Red Sea

The Red Sea Basin has the potential to provide viable metallic and nonmetallic natural resources. Because there are seven different countries bordering the economic zones, orderly exploration has not always been possible in the light of continuing disputes and lack of cooperation. Yet recent signs of economic cooperation between these countries lends an optimistic note to this discussion, particularly the United Nations Development Programme/World Bank hydrocarbon study project (O'Connor, 1992) and the Red Sea Commission, which was set up for the economic development of the Red Sea metalliferous sediments.

Hot Brines and Heavy Metal Deposits

The circulation of seawater through the still-hot oceanic crust within the Red Sea axial zone has produced hot brine pools underlain by metalliferous muds enriched in heavy metals (Degens and Ross, 1969; Backer and Schoell, 1972; Amann et al., 1973; Backer, 1975). The discovery of these areas was initially brought to the attention of the scientific community as a result of scattered measurements from research oceanographic ships in transit through the Red Sea (Bruneau et al., 1953; Neumann and McGill, 1962). Detailed measurements carried out by *Atlantis II* as part of the International Indian Ocean Expedition (1963–1964) reported temperatures of 25.6°C with a salinity of 43.2% at depths below 2000 m (Miller, 1964). When *Atlantis II* returned to the same area in 1971, analysis of cores from muds below the brine pool revealed extremely high concentrations of heavy metals (Miller et al., 1964). These discoveries led to optimistic estimates of precious metals (silver and gold) and provided an impetus for careful studies of the deposits. Definitive research was completed initially by the Woods Hole Oceanographic vessel *Chain* (1966) and the Preussig AG *R/V Valdiva* (1971–1972).These studies resulted in a symposium volume by Degens and Ross (1969) and in individual reports by Backer (1975), Schoell et al. (1974), and Amann et al. (1973). Contemporaneous with these studies, a cruise by *R/V Glomar Challenger* (Leg. 23B) was carried out in the central and southern Red Sea (Whitmarsh et al., 1974). These various research efforts resulted in a clear picture of the occurrence and distribution of the brines and associated metalliferous muds within the Red Sea (Ross, 1983).

More than 16 separate deeps with active brines, as well as inactive deeps with only hydrothermal sediments, have been located. New deeps with brines continue to be reported as more detailed oceanographic studies are carried out (Pautot et

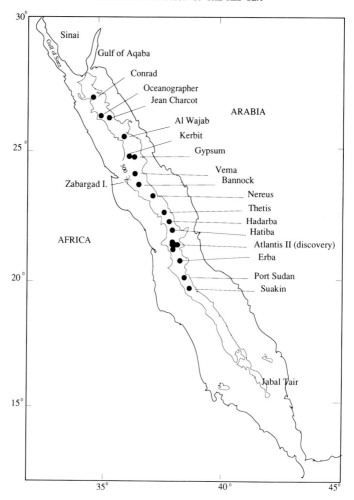

FIG. 8.1. Approximate locations of hot brine pools and deposits of heavy metals in the Red Sea; 500 m contour line shown.

al., 1984; Cochran et al., 1986) (Fig. 8.1). All of the hydrothermal occurrences are restricted to the axial part of the trough and appear to have been formed during the last 100,000 years (Degens and Ross, 1969). Present-day active brine pools such as the Atlantis II deep are overlain by waters of very high salinity and temperature as revealed by acoustic reflections and temperature profiles at depths of 2000 m (Fig. 8.2). Continuous temperature measurements with depth show distinct layers, with the upper layer at 43.3°C and the lower layer at 56.5°C (Ross, 1972). Repeated measurements from the same area with a time lapse of 4.3 years give similar temperature stratification and density values, but minor temperature variations indicate very slow convection or mixing between layers (Ross, 1972) (Fig. 8.2).

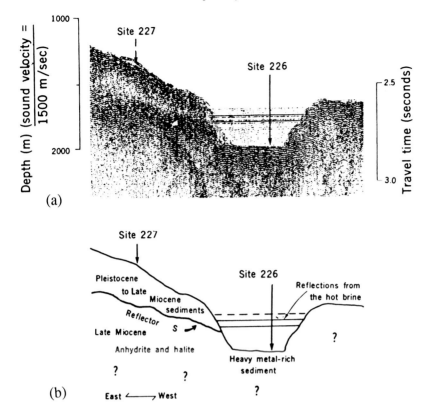

FIG. 8.2. (a) Seismic reflection profile and (b) interpretation of the Atlantis II deep where DSDP sites 227 and 226 were located (Ross et al., 1973).

The most extensively studied brine pool is the Atlantis II deep which may be used as a model for the other Red Sea brine pools. The early reports (Degens and Ross, 1969) show that these pools consist of many layers, representing density-stratification. Individual layers exhibit distinct temperature and salinity values (Zierenberg, 1990). The composition of the Atlantis II brine is compared on a spider diagram (Fig. 8.3) with normal seawater and with hydrothermal fluids emanating from active "black smokers" from the East Pacific rise and the Juan de Fuca Ridge. The similarity of these fluids is striking in that they are all saturated with respect to NaCl and are enriched in Na, K, Rb, and Ca when compared to seawater. Depletion of both Mg and SO_4 in the brines can be explained by the interaction of basalt-seawater with the underlying basalts (Mottl and Holland, 1978). These reactions also add large amounts of Cu, Zn, and Pb to the brines by leaching of the basalt, producing a hydrothermal fluid uncontaminated by the overlying seawater (Potorrf and Barnes, 1983). Variation in temperature of the brine pools have been observed by detailed hydrographic observations over a period of time, and indicate that they are still active (Ross, 1972; Schoell, 1976;

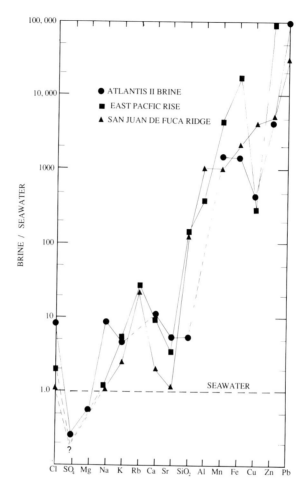

FIG. 8.3. Spider diagram comparison of chemical composition of seawater, (●) Atlantis II brine, (■) East Pacific (21°N) and (▲) Southern Juan de Fuca Ridge hydrothermal fluids. Data from Zierenberg (1990).

Schoell and Hartmann, 1978; Hartmann, 1980). These observations suggest that the brines are venting in various parts of the pools and must have different compositions. The brines rise vertically until they reach the interface between the lower and upper brines, where they spread laterally, mixing with the existing brine and eventually depositing their suspended base-metal sulfides (Zierenberg, 1990) (Fig. 8.4). Variations in the amount and intensity of these venting brines probably lead to development of distinctive stratigraphic facies, as observed through piston coring of the Atlantis II deep (Zierenberg, 1990).

These brines are considered to be derived from deep circulation of hydro-thermal fluids (initially seawater) within the underlying basalts at temperatures in excess of 330°C, and from a shallower fluid at about 250°C circulating within the evaporite–shale sequence to produce a brine rich in sulfate and low in reduced sulfur (Potorrf and Barnes, 1983) (Fig. 8.4). Analyses of the isotopes of strontium, sulfur, carbon, and oxygen in the metalliferous sediments in the brine pools indicate three major sources of the components that make up the brine: seawater, Miocene evaporites, and basalt formed in the axial zone (Zierenberg and Shanks, 1986, 1988). The mixture of these two brines probably took place below the present level of the brine pool and explains the disequilibrium between H_2S and SO_4 in the fluids where both pyrite and anhydrite are formed together. These hydrothermal brines derived from alteration and leaching of the evaporite and basalt sequences are much denser than seawater and will accumulate in graben bottoms within the active Red Sea rift zone. In contrast, the ridge-crest hydro-thermal brines are less dense than average seawater and form buoyant plumes resulting in a more dispersed deposit with sulfide-rich mounds near the vent areas.

The underlying metalliferous muds vary in thickness within the deposition basins of the deeps (3 to 20 m), and their estimated deposition rate of 1 mm per year is ten times greater than normal deep-ocean sedimentation. These sediments consist of variable mixtures of oxides, hydrosilicates, and sulfides, and their variability in composition indicates extreme changes in the sedimentary environment

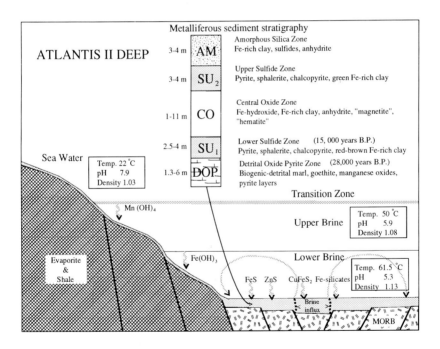

FIG. 8.4. Conditions in the Atlantis II brine pool, showing suggested geochemical processes of mineral precipitation and stratigraphy of the metalliferous sediments within the Atlantis II deep. Not drawn to scale, and derived from several sources (Bischoff, 1969; Potorrf and Barnes, 1983; Zierenberg, 1990).

(Ross and Degens, 1969; Potorrf and Barnes, 1983). The sediments are well
bedded as thin laminated layers whose colors range from black to white, with
yellow, blue, and red gradations from layer to layer. The grain size is within the
clay–silt range, but larger hyaloclastic basaltic fragments may be present. As
might be expected, the various facies of the muds vary widely in composition,
SiO_2 between 8 to 27%, $FeO + Fe_2O_3$ between 8 to 66% and Al_2O_3 usually less than
8%. These strange sedimentary compositions represent chemical precipitates
derived from hydrothermal brines that have widely variable concentrations due to
the mixing of fluids from the underlying basalts and the flanking evaporite–shale
sequence. Where piston cores and drilling have penetrated the metalliferous muds
they rest unconformably on newly formed basaltic hyaloclastites and pillow lavas
(Bischoff, 1969; Craig, 1969; Bignell et al., 1976) (Fig. 8.4).

A general basin-wide stratigraphy developed by Backer and Richter (1973)
indicates a period of approximately 28,000 years of hydrothermal activity (Fig.
8.4). The basal unit, called the detrital oxide pyrite (DOP) zone, consists mainly
of a biogenic–detrital marl that is depositional on the pillow lavas of the axial
spreading zone. Enriched zones of goethite and manganese oxides are interlayered
and increase upward. Also present are pyrite-bearing layers with high copper
content, whose sulfur isotope ratios indicate a biogenic origin for the sulfur.

The lower sulfide zone (SU_1) above the DOP zone consists of reddish-brown
iron-rich clays interlayered with blackish fine-grained sulfide-rich sediments
containing pyrite, sphalerite, and chalcopyrite. Backer and Richter (1973) consider
this unit to represent initiation of a permanent brine pool about 15,000 years ago
(Shanks and Bischoff, 1980).

The SU_1 zone grades upward into the central oxide (CO) zone, which contains
fine-grained to amorphous hydrated iron oxides and iron-rich clays. Discontinuous
horizons of sedimentary manganite and hematite are present, along with anhydrite-
rich layers near the top of this zone.

The second sulfide zone (SU_2) is quite similar to SU_1, consisting of iron-rich
clays which are predominantly green in color. In the southwestern part of the
Atlantis II basin the SU_2 zone shows disruption, with slumping and re-sedimentation.
The zone is capped by the amorphous silicate zone (AM), represented by unconsolidated admixture of iron-rich clay minerals and base-metal sulfides. Interstitial
brines in these sediments may be as high as 95 wt% but the general nature of the
AM zone is nearly identical to the SU_1 and SU_2 zones, indicating that the AM zone
is in the process of consolidation (Zierenberg, 1990).

The main controls influencing formation of the metalliferous sediments in the
Atlantis II deep are the supply of hydrothermal brines, variation in the suspended
material, and the tectonics of the spreading center. The high temperature, salinity,
and heavy metal content forms a sterile anoxic environment. The lack of all life
forms leads to finely laminated sediments undisturbed by bioturbation or by bac-
terial processes that lead to to sulfate oxidation or reduction (Zierenberg, 1990).
This anoxic brine environment forms a protective cover over the newly precipit-
ated sulfides from oxidation by the overlying oxygen-rich seawater (Fig. 8.4).

A comparison of the Atlantis II metalliferous sediments with ancient massive
sulfide deposits reveals a number of similarities. The often invoked idea that

these ancient massive sulfides can be related to circulating seawater that penetrates downward and leaches metals into hydrothermal systems which then discharge on the seafloor fits very nicely with the actual situation in the Atlanits II deep. In ancient massive sulfide deposits the formation of banded ore consisting of pyrite, chalcopyrite, sphalerite, and galena associated with chert, and iron-rich chert laterally grading away from the massive sulfide ore (Ohmoto and Skinner, 1983) is similar to the Atlantis II deep, where we observe the sulfide zones grad-ing upward into oxide and silica-rich zones.

Thus, the Red Sea metalliferous deposits resemble many volcanogenic massive sulfide deposits (Kuroko type, Cyprus type) in their tectonic setting, sulfide mineralogy, metal grade, sedimentary textures of the ores, and similarity in the hydrothermal ore fluids. The widespread occurrence of metal deposits in active ridge crests invites further comparison with the Red Sea brines and metalliferous sediments (Rona, 1980). Ridge crest massive sulfides developed from "black smokers," and the Red Sea metalliferous sediments develop from hydrothermal fluids formed by water–rock interaction of seawater at temperatures of about 350°C. However, the hydrothermal circulation within the evaporite–shale sequence in the Red Sea (Atlantis II deep) produces a very high-salinity, dense brine that collects in a submarine brine pool, producing a lower grade blanket deposit of much greater extent than that of the massive sulfide deposits with higher metal grades. Thus the ridge crest massive sulfide deposits require rapid sedimentary burial after their formation to protect them from continuing oxidation and dispersal by seawater (Zierenberg, 1990).

Unusually high levels of copper, zinc, silver, gold, lead, and iron in these sediments stimulated interest in their economic exploitation (Emery et al., 1969). An estimate for the southern part of the Atlantis II deep on a salt-free and dry basis indicates a volume of 150 to 200 million tons of "ore" containing 6.0% Zn, and 0.8% Cu: economically attractive at 1975 prices, given that ordinary mining costs of excavation and crushing, as well as smelting preparations, would not be required (Amann et al., 1973; Backer, 1975). A more recent estimate of the metal content based on dry, salt-free metalliferous sediments indicates 92 million tons with average grades of 2.1% Zn, 0.46% Cu, 41 g ton^{-1} Ag, 0.51 g ton^{-1} Au and 59 g ton^{-1} Co (Guney et al., 1988).

In 1978, it was reported that a Saudi–Sudanese Commission for the development of these metalliferous deposits had been organized, and, following pilot plant mining experiments, it was predicted that production would start in 1985, yielding between100 and150 thousand tons per annum with an estimated value of $2.5 to 3.0 billion over the next 15 to 20 years (Sardar, 1978; Nawab, 1984). The year after these optimistic projections were made, discovery of massive deep-sea sulfide ore deposits associated with hydrothermal vents within the active spread-ing centers of the East Pacific rise opened a new perspective on the exploitation of heavy metal deposits in the deep oceans (Francheteau et al., 1979). This new class of metallic deposits contain up to 29% Zn and 6% Cu, and it is possible that such deposits are distributed intermittently all along the spreading centers of the oceans (Rona, 1988). Comparison of potential mineral resources within the sea-floor spreading centers from various tectonic settings such as the Red Sea metalliferous muds and massive sulfides of fast spreading ridges needs to be made (Edmond et al., 1982; Haymon and Macdonald, 1985; Rona, 1988; Haymon, 1989). Nearly ten years later the Saudi–Sudanese Red Sea Joint Commission issued an important report outlining the toxic dangers in mining metalliferous muds from the Atlantis II deep, and now it is not clear if the environmental risk is worth the economic exploitation (Karbe, 1987). The absence of bacteria and the very high toxic metal content of the brines makes their mining extremely dangerous, and it may disturb the proper balance in the Red Sea itself (Edwards and Herad, 1987).

Potential for Petroleum

The sedimentary history of the Red Sea basin started in the Oligocene, with lacustrine sediments and continental red beds interlayered with bimodal volcanics grading upward into infra-evaporite deep marine muds and sands to an enormous thickness (4 to 5 km) of Miocene evaporites extending from the Gulf of Suez to just north of the straights of Bab El Mandeb. These salts are in turn overlain by marine Pliocene–Pleistocene pelagic sediments in the central portion of the basin, interdigitating with coarse clastics derived from the uplifted margins of the basin.

The potential for oil is therefore considerable, bearing in mind that the underlying pre-Miocene sediments are petroleum source rocks and that overlying porous horizons allow the migration of petroleum to traps under the salt or along the flanks of salt diapers. The intensity of oil exploration in this area is diminished by the presence of much larger oil fields in eastern Arabia, where proven reserves are much less expensive to develop. There has been a prevalent "oil-industry" opinion concerning the Red Sea Basin; that it is generally gas-prone as a result of high heat flow. This opinion has been greatly modified since the publication of the UN Development Programme/World Bank Red Sea–Gulf of Aden Regional Hydrocarbons Study Project, in two special issues of the *Journal of Petroleum Geology* (Beydoun, 1989; Beydoun and Sikander, 1992a, b; O'Connor, 1992) (Fig. 8.5, Table 8.1). The following discussion is based on these sources.

Gulf of Suez

The Gulf of Suez, with nearly 800 exploratory wells, is the only part of the Red Sea Basin where oil production has been established. These wells, along with geophysical information, provide a database that has facilitated greater understanding of the geological history of the area (Fig. 8.6; see also Fig. 2.7). The Gulf of Suez is unique in that it is a failed continental rift that was ended by the Dead Sea fault system in the Middle Miocene. The termination of this active continental extension and loss of the heat source produced by the rising asthenosphere produced a unique situation for petroleum maturation within the Suez basin.

The pre-rift rocks underlying the Gulf of Suez are Precambrian crystalline basement overlain by Nubian sandstones of undetermined age. Unconformably overlying the Nubian sandstones are Late Cretaceous marine deposits (about 800 m), deposited below the southern margin of the Tethyan ocean and considered

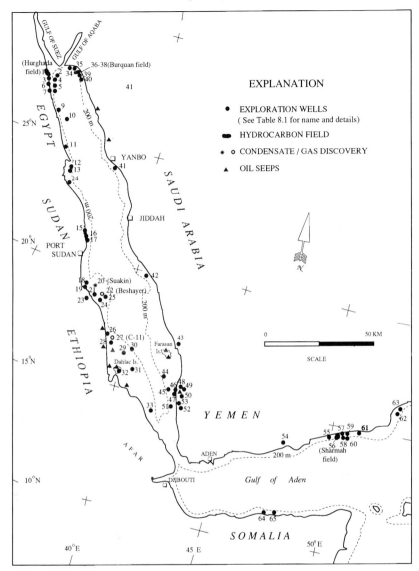

FIG. 8.5. Generalized map of exploration oil well locations in the Red Sea and Gulf of Aden. After Beydoun (1989). See Table 8.1 for a description of each well.

to be the main source rock for the Gulf of Suez oil (Helmy, 1990). A section of Paleocene–Eocene calcareous marine shales (60 to 300 m) unconformably overlie the Tethyan sediments. A marked angular unconformity separates the syn-rift Lower to Middle Miocene clastics. Mafic flows and intrusives within these early rifting continental clastics mark the beginning of rifting volcanic activity.

Tertiary sediments mark the beginning of rifting and crustal thinning. The overlying evaporites form an effective seal, and range in thickness from 230 to

TABLE 8.1. Petroleum exploration wells in the Red Sea and Gulf of Aden. The numbers correspond to those in Fig. 8.5, which shows relative location. From Beydoun (1989).

Country		Well name (see Fig. 8.5)	Total depth (m)	Operator, year completed	Hydrocarbon shows or test results	Remarks
Egypt	1	Hurghada field		GPC Producing 1915–1969	Depleted oil field	22° – 30° API oil from Miocene dolomite and basal Miocene and Nubia (Cretaceous) sands, 41MM brls oil produced
	2	RSO-T 95-1	1951	Esso 1977	Dry	
	3	Um Agawish-1	1678	Mobil 1979	Dry	
	4	RSO-X 94-1	2900	Esso 1981	Shows oil and gas	Thin sands within evaporite group and Infra-evaporites
	5	RSO-Z 95-1	3650	Esso 1977	Shows	Thin sands within Evaporite group
	6	Ras Abu Soma-1	1321	Canadian ?	11 brls oil 29°–32° API	Swabbed on production test from 1st. (basal Belyim M. Miocene) at ca. 800 m depth
	7	RSO-B 95-1	4022	Esso 1975	Shows	Thin sands within evaporite group
	8	RSO-B 96-1	4250	Esso 1981	Shows	Sandstones in basal Miocene
	9	Qusair A-IX	5038	Phillips 1977	Dry	
	10	Qusair B-IX	4214	Phillips 1977	Dry	
	11	RA West-1 (RBO-Z 108-1)	1746	Union 1975	Dry	
	12	Abu Madd-1 (RBO-G 112-1)	1627	Union 1975	Shows	?Infra-evaporites. Abnormally high pressures at ca. 1600 m
	13	Mikawa-1	3362	Total 1985	Dry	

In addition, a number of shallow wells have been drilled near Dishet el Dabaa and Giftun Islands south and southeast of Hurghada field, some of which are of post-World War II vintage.

Country		Well name	Total depth (m)	Operator, year completed	Hydrocarbon shows or test results	Remarks
Sudan	14	Halaib-1	3596	Texas Eastern 1982	Dry	
	15	Dungunab-1	1615	Agip 1962	Dry	
	16	Abu Shagra	2293	Agip 1962	Dry	
	17	Maghersum-1	2000	Agip 1962	Dry	
	18	Durwara-1	2900	Agip 1961	?Shows. Suspended	
	19	Durwara-2	4152	Agip 1963	Gas shows	Non-commercial
	20	Suakin-1	2745	Chevron 1976	1160 b/d 52° API condensate and 6.9 MM cu. ft/d	? formation

Sudan	21	Bashayer-1A	2787	Chevron	1976	9.5 MM cu. ft/d	Sandstone in supra-evaporite (Zeit formation)
	22	Bashayer-2	2599	Total	1981	Gas shows	Non-commercial
	23	Tokar-1(Marafit-1)	3119	Agip	1963	Dry	
	24	Digna-1	2194	Union Texas	1982	Dry	
	25	South Suakin-1	3713	Chevron	1977	Dry	
Ethiopia	26	J-1	3137	General American	1973	Dry	
	27	C-1	3010	Mobil	1970	Gas blow out	Carbonate intercalation within basal salt.(Evaporite Group) Ran wild for 55 days
	28	MN-1	2870	General	1973	Dry	
	29	Amber-1	3557	Mobil	1966	Dry	
	30	B-1	2966	Mobil	1969	Dry	
	31	Dhunishub-1	3867	Gulf	1966	Dry	
	32	Secca Fawn-1	3363	Gulf	1969	Gas shows	In carbonate near TD at base evaporite group and show at ca. 2135 m
	33	Thio-1	3119	Shell	1977	Dry	

In addition, a number of wells were drilled on Dahlak Islands by Agip in 1940, mostly a few hundred meters in depth, but with two reaching depths over 2000 m (Suri-7, 2503 m; Adal-2, 2475 m); oil shows were encountered.

Saudi Arabia	34	Al Kurmah-1	3122	Auxerap	1970	Dry	
	35	Rayaman-1	3810	Tenneco	1975	Dry	
	36	Barqan-1	2900	Auxerap	1969	Gas and condensate flow	Four gas-condensate sands in infra-evaporite group (Miocene)
	37	Barqan-2	2786	Auxerap	1969	Gas and condensate flow	Estimated absolute open flow 100 MM cu. ft/d. 11.6 MM cu. ft and 650 b/d condensate from sands between 1932 m and 1977 m in Infra-Evaporites (Gharandal group) Karem-Rudeis formation
	38	Barqan-3	3227	Tenneco	1971	41.8° API oil	In sands of Globigerinal Marl Group (Gharandl group) or Infra-Evaporite Group
	39	Yuba-1	2884	Auxerap	1969	Dry	
	40	An Numan-1	2100	Tenneco	1976	Dry	

continued

TABLE 8.1. *continued*

		Well	Depth	Operator	Year	Result	Comments
Saudi Arabia	41	Badr-1	3347	Sun	1972	Dry	High pressure and temperature (175°C)
	42	Ghawwas-1	3466	Sun	1971	Water blow out at 3447m.	
	43	Mansiyah-1	3931	Auxerap	1967	Shows	In thin sands within the main Evaporite Group at 2500 m

In addition, a number of shallow holes were drilled by PD (WA) between 1933 and 1939 on the Farasan Islands near the Ras Hassis oil seeps and elsewhere.

		Well	Depth	Operator	Year	Result	Comments
North Yemen	44	Al Meethag-2	2042	Hunt	1987	Dry	
	45	Al Meethag-1	1836	Hunt	1987	Dry	
	46	Salif-1	1524	Mecom	1961	Dry	Stratigraphic well
	47	Salif-2	2222	Mecom	1961	Shows?	
	48	Abbas-1	3414	Shell	1981	Gas shows	Non-commercial in carbonate within Evaporite unit
	49	Zaidiya-1	3017	Mecom	1962	Some oil on DST	At about 213 m. Non-commercial
	50	Al Auch-1	2812	Shell	1980	Minor shows	
	51	Kathib-1	2459	Shell	1976	Dry	
	52	Hodeida-1	1729	Mecom	1962	Dry	Mechanical problems
	53	Hodeida-2	2732	Mecom	1963	Dry	Abandoned (political factors)
South Yemen	54	Balhalf-1	3947	Elf	1986	Dry	In onshore northwest–southeast graben
	55	Hami-1	2424	Agip	1981	Dry	
	56	Sharmah-1	2984	Agip	1982	3700 b/d 43° API oil	Middle Eocene carbonates at about 2100 m. Oil shows in Oligocene
	57	Ghaswah-1	3349	Agip	1979	Dry	
	58	Ghaswah-1	2845	Agip	1983	Dry	
	59	Qusayr-1	2190	Agip	1982	Dry	
	60	Sarar-1	3483	Agip	1981	Oil shows	In Cretaceous sands
	61	Masila-1	4397	Agip	1979	Dry	
	62	Al Fatk-1	4300	Agip	1981	Dry	
	63	Al Faydany	1656	Agip	1982	Dry	
Somalia	64	Bandar Harshau-1	2504	Shell	1985	Dry	
	65	Dab Qua-1	3159	Shell	1984	Dry	

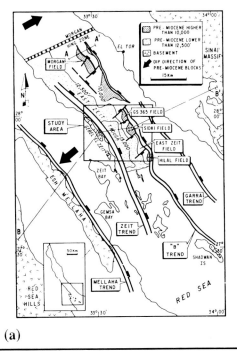

(a)

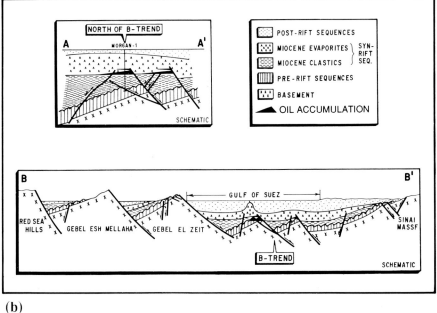

(b)

FIG. 8.6. (a) General structural map showing pre-Miocene trends (contour interval 2500 feet). (b) Cross-sections of the southern Gulf of Suez, locations shown in (a). A–A' trend in the north and B–B' in the south (Helmy, 1990).

2400 m. Overlying the evaporite sequence are unconsolidated and interbedded sands, clays, and marls (150 to 600 m) that are Pliocene to Recent in age and contain a benthic fauna characteristic of the Indian Ocean (Miller and Barakat, 1988; Helmy, 1990). This important unconformity is recognized as the S-reflector throughout the Red Sea Basin (Phillips and Ross, 1970).

The Gulf of Suez is dominated by North–northwest normal faults which produce three distinct dip provinces, depending on the fault geometry. These faults produce traps in the syn-rift sediments which are sealed by the evaporite (Helmy, 1990). It is predicted that continued modeling of the stratigraphy and structure of sub-surface data combined with exploratory drilling will lead to further discoveries (Ott d'Estevou et al., 1987; Evans, 1988; Helmy, 1990).

Red Sea

Approximately 53 exploration wells have been drilled by various oil companies in the Red Sea as of 1992 resulting in four dry gas discoveries (Beydoun, 1989) (Fig. 8.5 and Table 8.1). The UN Development Programme study group consisted of oil companies with previous exploration experience and resources in addition to a close working relation with the countries surrounding the Red Sea. Speciality groups studied the source rock geochemistry (Institut Francais du Petrole), biozonation and chronostratigraphy (Simon-Robertson), and tectonic framework (Geophysical Institute of Hamburg University). The multinational basin study was a success and has been used as a model for subsequent explorations.

Pre-rift source rocks important to oil generation are present in the basin. As shown earlier, the area centered around the Red Sea Basin during the Paleozoic–Mesozoic was exposed as a low-lying area with laterite forming on the surface in the Cretaceous–Eocene. Onlap of marine transgressions from the Tehtyan ocean produced marine carbonate and clastic deposits Cenomanian to Paleocene in age extending as far south as 20°N on the Sudan coast (Mukawar formation) and at Jiddah (Shumaysi formation) on the Saudi Arabian coast. These rocks have been identified as source rocks with significant organic content in the Gulf of Suez and in the Quesir area on the Egyptian side within the phosphate-rich Duwi formation, which may have thicker and deeper oil shale offshore (Barnard et al., 1992).

Jurassic clastic sediments present in Yemen, Ethiopia and southern Saudi Arabia are thought to be marginal-marine to continental sediments with good reservoir potential (Mitchell et al., 1992). Resting unconformably above the Jurassic clastic sequence is a thin series of shallow marine limestones (Amaran formation, Yemen and Saudi Arabia; Anatalo formation, Ethiopia) that has not been penetrated by exploratory wells in the southern Red Sea. These lithographic limestones (up to 500 m) may be potential source rocks, as age equivalent limestones in eastern Yemen are considered to be the source rock for new oil-fields in the Hadramut area (Haitham and Nani, 1990; Paul, 1990). The pre-rift source beds are widespread and discontinuous, but may be significant as source beds throughout the Red Sea, particularly if it is assumed that there has been extensive extension of the continental crust on both sides (Fig. 8.7).

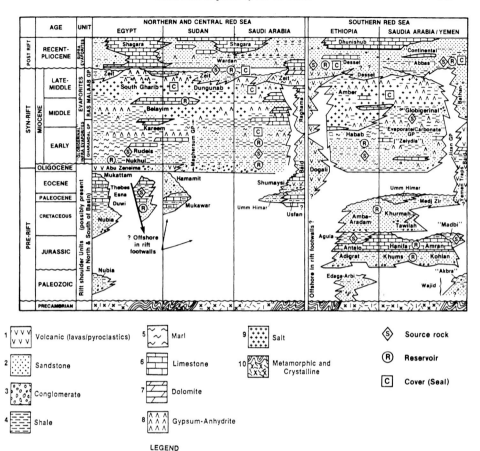

F<small>IG</small>. 8.7. Generalized stratigraphic correlation chart for the Red Sea Basin, indicating units having petroleum source potential, reservoir capacity, and cap rock characteristics. After Beydoun (1989).

Initial rift continental red beds and volcanics range from Eocene through Oligocene and characteristically consist of coarse alluvial continental red beds interlayered with bimodal volcanic and lacustrine deposits. Nearly all of these deposits thin towards the axial part of the Red Sea. They can be found in the Dogali formation in Ethiopia and in the Hamamit formation in Sudan. The petroleum source potential for these rocks is very low because of the continental alluvial origin and lack of organic material. However, the reservoir potential of such rocks captured in half grabens produced by Oligocene block faulting is important (Mitchell et al., 1992). Reports of undated lacustrine deposits in some of the deep wells indicate that the onshore Baid formation lacustrine sediments may be widespread during the initial rift phase and could possibly be a source rock (Crossley et al., 1992). The lack of good biostratigraphic correlations for these lacustrine sediments leaves the question of their source capabilities open to further studies (Fig. 8.7).

Following the initiation of rifting in the Oligocene a marine connection was established to the north through the Gulf of Suez, with the formation of the Middle Miocene clastic wedge (infra-evaporite or syn-rift sediments). These marine mudstones and interlayered sands (Habab formation, Globigerina marls, Maghersum formation) reach a thickness of 2000 m and represent deep marine conditions similar to the Lower Miocene formations Kareem and Rudies in the Gulf of Suez, both of which are considered to have good oil potential (Crossley et al., 1992). Exploratory wells in the southern Red Sea rarely penetrate these marine clastics, so their characteristics in the central part of the southern Red Sea are unknown (Fig. 8.7).

The regression of the Red Sea began in the Middle Miocene, leaving mixed marine and minor evaporite sequences, which finally resulted in thick evaporite sections being deposited rapidly up to the Pliocene. The UN Development Programme study group concluded that conditions for organic preservation existed during the salinity crisis, with frequent indications of important oil potential (Barnard et al., 1992). Interlayered and reworked alluvial sands in the evaporite sequence provide potential reservoirs.

At the beginning of the Early Pliocene an abrupt change to deep-marine deposition with normal salinity is recorded throughout the basin as the S-reflector (see Fig. 2.10) (supra-evaporites, post-rift sediments). Accepted knowledge is that this seawater accessed through the Straits of Bab el Mandeb. These sediments vary from deep to shallow marine, with marginal clastic sediments derived from the uplifted flanks of the Red Sea. Crossley et al. (1992) suggest that the aprupt flooding of the shallow-water Red Sea evaporite basin may have released large quantities of terrestrial-sourced nutrients, giving rise to high productivity with sediment petroleum source potential. The local influx of sand from catchment areas around the Red Sea (Tokar delta) may give rise to syn-salt clastic sedimentation or post-salt halokinetic deformation by sediment loading of the salt (Fig. 8.8).

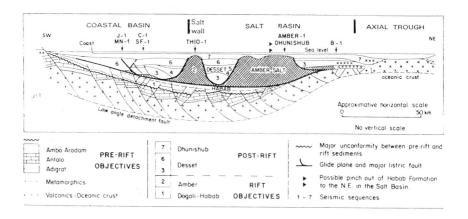

FIG. 8.8. Schematic structural section of the western margin of the southern Red Sea, extending from the Ethiopian coast and showing the approximate location of exploratory wells. From Savoyat and Balcha (1989).

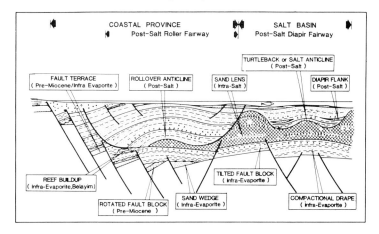

Fig. 8.9. Generalized cross-section illustrating hydrocarbon "play" systems common in the Red Sea. From Mitchell et al. (1992).

Mitchell has brought to our attention two hydrocarbon systems (plays) within the Red Sea: (1) Pre-evaporite fault block rotation similar to that in the Gulf of Suez is present in the northern Red Sea; and (2) post-evaporite, mostly in the southern Red Sea, where increasing heat flow and high sediment loads generate overmature pre-evaporite sequences almost everywhere except along basin margins. Along the margins the oil maturation and halokinetic post-salt structure combine to form attractive hydrocarbon "plays" (Fig. 8.9).

Beydoun and Sikander (1992a) summarize the hydrocarbon play types as follows:

> Of the 53 or so exploration wells drilled in the Red Sea Basin (the overwhelming majority of which have been located offshore) there have been three undeveloped hydrocarbon discoveries (Barqan in the Saudi Arabian offshore with light oil, gas and condensate; Suakin in the Tokar delta area of offshore Sudan with gas and condensate; Beshayir, close to Suakin with gas) and one unassessed gas blowout (C-1, offshore Eritrea or Ethiopia). In addition, the Hurghada depleted oil field is located in the Egyptian coastal plain at the junction of the Red Sea proper with the Gulf of Suez: this has produced substantial amounts of 22-30° API oil between 1915 and 1969. Moreover, many of the wells encountered oil and gas shows during drilling whereas many surface oil seepages are documented from various parts of the Red Sea Basin, particularly from the Farasan and Dahlak islands in the South (Beydoun, 1989).
>
> Attractive pre-rift sources and reservoirs deposited in Arabian shelf settings are present at outcrops in the Egyptian sector (Upper Cretaceous) and along the southern Red margins (Upper Jurassic), but these may have been removed from the crests of rotated fault blocks which constitute the main play targets, or be overmature for oil where they may be preserved in half graben lows; in the southernmost Red Sea area, however, they are likely to be preserved at moderate depth and be prospective for oil (Mitchell et al., 1992).

The main Red Sea plays are for Miocene syn-rift targets, both pre-and post-salt. In the pre-salt, good source potential exists in top Lower to Middle Miocene levels in the Belayim and Karim formations of Egypt, but in the central and southern Red Sea they are probably overmature for oil except in marginal locations and in the southernmost sector near Bab el Mandeb outside the main salt basin. The Lower Miocene Rudies formation has good source potential in the Egyptian sector but becomes only fair to poor in quality further south and is also likely to be overmature for oil to the south except in marginal settings. On the Saudi Arabian side, excellent Lower to Middle Miocene source rocks ("Globigerinal Marls") are present in the north and have most probably generated the Barqan hydrocarbons; these source sediments probably extend elsewhere along the eastern side but the lack of well control precludes precise delineation. Reservoirs are provided on a regional basis by coastal sandstones of littoral or deltaic environmental and prograding clastic wedges as well as from erosion of irregular fault block surfaces; shoal and reefal carbonates on the crests of blocks are known from Egypt and Sudan. Seals are provided by mudstones or interlayered anhydrites. Hydrocarbons would be trapped in rotated and/or tilted fault blocks, fault terraces, sand wedges, compactional drape folds and carbonate buildups (Mitchell et al., 1992).

Source-reservoir and seal facies in syn-rift and, to a lesser extent, in post-rift (drift) sequences, however, exhibit rapid lateral sediment variability as a consequence of quick deposition in the (rapidly) subsiding rift basins and of variation in the influx quality and quantity of sediment fill, in contrast to the much wider areal continuity of pre-rift facies. This, naturally, may place limitations on areal hydrocarbon pool sizes but increases the chances of pure stratigraphic trap accumulations and of greater vertical extent of multi-pool accumulations as a result of cyclicity of sedimentation. During the main salt deposition time in the southern Red Sea there is evidence of greater subsidence on the Ethiopian side as well as indications that the Red Sea axial zone was positive at that time dividing the southern area into two salt basins (Mitchell et al., 1992).

For the younger syn-rift sequences, including the salt and post-salt, thin high quality dominantly oil-prone source levels occur within the the evaporite sequence of the Middle–Upper Miocene South Gharib formation (and lateral equivalent main salt formations) and the overlying Zeit formation laid down during low salt production and periods of severe drawdown (Crossley et al., 1992). Halokinetically controlled depocenters accented during the Pliocene contain the main source kitchens of the post-salt Zeit and possibly also the lower parts of the Warden (= Desset and Abbas) formations and are more likely to be mature and prolific in the central and southern Red Sea and perhaps mature in the Egyptian sector. The seismic data show that listric faults in the post-salt sole out at the base of the salt (Mitchell et al., 1992). In the post-rift succession, good oil-prone source rocks with high TOCs were identified from the lower Warden (Pliocene) sequence of Hurghada in the Egyptian sector (Barnard et al., 1992) and may be present in inter-diapir lows elsewhere in the basin, although probably immature on the flanks except in the southern part. These would charge North Sea-like alluvial fan reservoirs debouched into these halokinetically induced depressions making exploration of the synclinal depocenters of this younger sequence a desirable objective. Main reservoirs for the post-salt section are provided dominantly by sands of deltaic littoral/intertidal or alluvial fan nature near the main sediment influx points. Sand lenses within the evaporites and the post-salt and post-rift reservoir sands are best developed further into the basin with axial distribution northwards from the Bab el Mandeb in the Early Pliocene. The evaporites themselves provide the principal seals and interlayered mudstones in the Zeit and Warden formations constitute additional seals. Traps are provided by roll-over anticlines, turtlebacks and diapir flank wedges and by sand lenses in the intra-salt (Mitchell et al., 1992).

The pre-rift and syn-rift pre-salt targets in rotated fault blocks similar to the Gulf of Suez form attractive plays in the northern Red Sea where geothermal gradients are moderate and comparable with those of the petroliferous Gulf of Suez. Support for this is provided by the undeveloped Bargan discovery in the northern Saudi Arabian sector. Thermal modeling indicates that in general the Miocene pre-salt succession reached the top of the main oil generation window about 10 Ma. Southwards, increasing heat flow and highest sediment load especially at sediment influx points on the African side, generally result in overmaturity of the pre-salt sequence for oil except along basin margins and perhaps in the platformic Bab el Mandeb region outside the main salt basin. Intra-salt plays seemingly have only limited potential but can be of local significance (e.g. the C-1 gas blowout in offshore Ethiopia). The generally higher heat flow to the south, together with halokinetic structuring of the post-salt Miocene (and Pliocene) successions, provide a number of attractive plays for oil as well as for gas and condensate, with support for this provided by the undeveloped Suakin and Bashayer discoveries in offshore Sudan and by the many surface oil seepages in the southern Red Sea. Thermal modelling suggests that for the post-salt Miocene succession the top of the main oil generating window was reached about 5 Ma in this region (Mitchell et al., 1992). Substantial volumes of recoverable hydrocarbons can be expected from this sparsely explored basin, particularly from the virtually unexplored central part of the Arabian side.

Epilogue

It must be obvious that this book has not dealt with all aspects of Red Sea earth science research. Our knowledge of the processes active during the formation of the Red Sea have greatly improved, and the intent is to integrate geological and geophysical studies. Lacking overall expertise in these studies, I have placed more emphasis on the volcanic and sedimentary histories.

The early manifestation of rifting is marked by continental stretching accompanied by large-scale invasion of the extended crust by tholeiitic magmas, either by intrusions or underplating. The first stages of rifting are marked by basin subsidence, with initial deposition of terrestrial sediments derived from uplift along the flanks of the basin. No geologic evidence has been found to support the existence of a large dome centered in the Afar Depression during the early stages of rifting that can be considered an initial hot spot. The Miocene salinity crisis of the Mediterranean area extended throughout the Red Sea Basin and the brines were fed from the western Mediterranean Sea into the Gulf of Suez. Normal block faulting preserved in rare exposures along the Red Sea margins supports the idea of continental stretching in the Miocene–Oligocene stages of rifting. The extent of this stretching is debatable because the thick evaporite sequence in the basin obscures the basement. Geophysical refraction lines across the margins of the Red Sea clearly reveal the thinning of continental crust on both sides of the basin. Interpretation of the magnetic and gravity measurements of the basin flanks along with the seismic data have produced numerous interpretations. One interpretation supports the presence of oceanic crust from shoreline to shoreline, whereas the opposite interpretation includes large increments of stretching, with oceanic crust confined only to the axial trough. Over the last twenty years I have wavered between these two extreme viewpoints. However, I now believe that the crust under the thick evaporite section in the Red Sea consists of extended Precambrian crust invaded by tholeiitic intrusives. Nearly everyone agrees that in the Pliocene change of plate motion initiated development of oceanic crust within the axial trough, but this is confined to the central portion of the Red Sea and appears to be propagating to the north and to the south. Sea-floor spreading was initiated earlier in the Gulf of Aden (Miocene) and is now propagating westward towards the Gulf of Tadjura. These propagating rifts are extending towards the postulated Afar hot spot, exactly the opposite of what would be expected if the Afar Depression represents the center of a large permanent hot spot! Passive rifting appears to have initiated early rifting by movement of the Arabian plate away from the African plate, and modern tomography along the Red Sea shows a mixture of hot spot activity and passive rifting. Continued volcanic activity within the Afar Depression may be a function of a local hot spot, but there is no compelling evidence that such a hot spot initiated continental breakup in the Miocene–Oligocene. Movement along the Dead Sea fault marks the beginning of a slow rotation of Arabia away from Africa. Movement of the Arabian plate and formation of new oceanic crust continues up to the present, satisfying kinematic schemes of plate movement. Evidence of extension and mafic volcanism in the Oligocene–Miocene indicates that the edges of the Arabian and African plates were not rigid, so shoreline matching that far back is unrealistic.

Solution of the early geologic history of the Red Sea requires much more work before final models can be realistically applied. The UN Development Programme/World Bank study of the hydrocarbon potential of the Red Sea–Gulf of Aden has gathered together and digitized much of the data on oil exploration. This information is now available in the Cairo Work Station (O'Connor, 1992). There is a need to compile a tectonic and geologic map of the Red Sea Basin at a scale of 1:500,000 using the stratigraphic information developed by the UN Development Programme/World Bank study group (Beydoun and Sikander, 1992b). Further seismic refraction profiling in the southern Red Sea is needed to establish the nature of the basement rocks. Drilling several deep scientific holes within the marginal parts of the Red Sea could produce new evidence on the nature of the crust underlying the basin. Most exploratory wells extended only a short distance below the evaporite section. Continued oceanographic studies, particularly in the southern Red Sea, are needed to match the more recent oceanographic work in the northern Red Sea.

All modern geologic indicators display evidence that the Red Sea will continue to grow by a combination of faulting and volcanic activity. Knowing this, it is reasonable to predict that future seismic and volcanic events will take place. Efforts should be made to prepare for potential disasters as well as to put in place systems to predict the nature and location of such hazards.

The Red Sea is a special laboratory for the study of rifted basins and contains a unique ecosystem within its waters. The unparalleled nature of this area should be given consideration before further exploitation of the heavy metal deposits in the axial trough and petroleum within the basin sediments is conducted.

A future scientific symposium, such as that sponsored earlier by the UN Development Programme/World Bank petroleum study group, involving the countries surrounding the Red Sea Basin could provide an additional incentive for making an inventory of the Red Sea scientific knowledge. A meeting such as this could consider problems of pollution and economic exploitation and add special parameters to global earth systems analyses. Developing cooperative studies with participation of scientists from all parts of the world under the sponsorship of the International Geological Cooperative Project (IGCP) could also provide a focus for research efforts within a planned scientific framework. This particular book could not have been attempted without such long-term close cooperation with scientists from countries bounding the Red Sea.

References

Abel-Gawad, M. (1970) Interpretation of satellite photographs of the Red Sea and Gulf of Aden. *Phil. Trans. R. Soc. Ser. A* **267**, 23.

Aki, K., Christofferson, A., and Husebye, E.S. (1977) Determination of the three-dimensional seismic structure of the lithosphere. *J. Geophys. Res.* **82**, 277.

Allan, T.D., (1970) Magnetic and gravity fields over the Red Sea. *Phil. Trans. R. Soc. Ser. A* **267**, 153.

Allan, T.D. and Pisani, M. (1966) Gravity and magnetic measurements in the Red Sea. *Canadian Geological Survey Paper (The world rift system)*, nos. 62-63.

Allan, T.D., Charnock, H., and Morelli, C. (1964) Magnetic, gravity, and depth surveys in the Mediterranean and Red Seas. *Nature* **204**, 1245.

Allegre, C.J. (1987) Isotope geodynamics. *Earth Planet. Sci. Lett.* **86**, 175.

Almond, D.C. (1986a) Geological evolution of the Afro-Arabian dome. *Tectonophysics* **131**, 301.

Almond, D.C.,(1986b) The relationship of Mesozoic–Cainozoic volcanicity to tectonics in the Afro-Arabian dome. *J. Volcanol. Geotherm. Res.* **28**, 225.

Almond, D.C., Kheir, O.M., and Poole, S. (1984) Alkaline basalt volcanism in northeastern Sudan: a comparison of the Bayuda and Gedaref areas. *J. Afr. Earth Sci.* **2**, 233.

Al-Shanti, A.M.S., (1966) Oolitic iron ore deposits in Wadi Fatima between Jiddah and Mecca. *Miner. Res. Bull.* **2,** Saudi Arabian Directorate General

Altherr, R. (1992) The Afro-Arabian rift system. *Tectonophysics* **204**, 1.

Altherr, R., Henjes-Knust, N.B.W., and Baumann, A. (1990) Asthenosphere versus lithosphere as possible sources for basaltic magmas erupted during formation of the Red Sea: constraints from Sr, Pb, and Nd isotopes. *Earth Planet. Sci. Lett.* **96**, 269.

Altherr, R., Henjes-Kunst, F., Puchelt, H., Baumann, A. (1988) Volcanic activity in the Red Sea axial – evidence for a large mantle diapir? *Tectonophysics* **150**, 121.

Amann, H., Backer, H., and Blisserbach, E. (1973) Metalliferous muds of the marine environment. *Offshore Tech. Conf., Dallas* **OTC 1759**, p.345.

Anderson, D.L. (1990) Geophysics of the continental mantle: an historical perspective. In: *Continental Mantle*. Oxford: Clarendon Press, p. 1.

Arkell, W.J. (1951) Origin of the Red Sea graben. *Geol. Mag.* **88**, 70.

Arno, V., Bakashwin, M.A., Bakor, A.Y., Barberi, F., Basahel, A., Di Paola, G.M., Ferrara, G., Gazzaz, M.A., Gilliani, A., Heikel, M., Marinelli, G., Nassief, A.O., Rosi, M., and Santacroce, R. (1980) Recent basic volcanism along the Red Sea Coast: the Al Birk lava field in Saudi Arabia. In: B. Zanettin (ed.) *Geodynamic Evolution of the Afro-Arabian Rift System*, 1979. **47**, 645. Rome: *Atti Dei Convegni Lincei.*

Avedik, F., Geli, L., Gaulier, J.M., and Le Formal, J.P. (1988) Results from three refraction profiles in the northern Red Sea (above 25° N) recorded with an ocean bottom vertical seismic array. *Tectonophyisics* **153**, 89.

Backer, H. (1975) Exploration of the Red Sea and Gulf of Aden during the *M.S. VALDIVA* cruises "Erzschlamme A" and "Erzschlamme B". *Geol. Jber. D.* **13**, 1.

Backer, H., and Richter, H. (1973) Die rezente hydrothermal-sedimentare Lagerstatte Atlantis-II-Tief in Roten Meer. *Geol. Rdsch.* **62**, 697.

Backer, H., and Schoell, M. (1972) New deeps with brines and metalliferous sediments in the Red Sea. *Nat. Phys. Sci.* **240**, 153.

Backer, H., Clin, M., and Lange, K. (1973) Tectonics in the gulf of Tadjura. *Mar. Geo.* **15**, 309.

Backer, H., Lange, K., and Richter, H. (1975) Morphology of the Red Sea central graben between Subai Islands and Abul Kizan. *Geol. Jber. D.* **13**, 79.

Baker, P.E., Brosset, R., Gass, I.G., and Neary, C.R. (1973) Jebel Abyad: a recent alkalic volcanic complex in western Saudi Arabia. *Lithos.* **6**, 291.

Baldridge, W.S., Eyal, Y., Bartov, Y., Steinitz, G., and Eyal, M. (1991) Miocene magmatism of Sinai related to the opening of the Red Sea. *Tectonophysics* **197**, 181.

Barberi, F., and Varet, J. (1975) Recent volcanic units of Afar and their structural significance (In a summary). *Afar Depression of Ethiopia.* Stuttgart: Schweizbart, 174.

Barberi, F., Boprsi, S., Ferrara, G., Marinelli, G., and Varet. J. (1970) Relations between tectonics and magmatology in the northern Danakil depression (Ethiopia). *Phil Trans. R. Soc. Ser. A* **267**, 293.

Barberi, F., Giglia, G., Martinelli, G., Santacroce, R., Tazieff, H., and Varet, J. (1971) Carte geologique du Nord de l'Afar, scale 1:500,000. La Celle–St Cloud, CNR–CNRS.

Barberi, F., Tazieff, H., and Varet, J. (1972) Volcanism in the Afar depression: its tectonic and magmatic significance. *Tectonophysics* **15**, 19.

Barberi, F., Bonatti, E., Marinelli, G., and Varet, J. (1974a) Transverse tectonics during the split of a continent: data from the Afar rift. *Tectonophysics* **23**, 17.

Barberi, F., Santacroce, R., and Varet, J. (1974b) Silicic peralkaline volcanic rocks of the Afar depression (Ethiopia). *Bull. Volcano.* **38**, 755.

Barberi, F., Capaldi, G., Gasperini, G., Marinelli, G., Santacroce, R., Scandone, M., Treuil, M., & Varet, J. (1980) Recent basaltic volcanism of Jordan and its implications on the geodynamic history of the Dead Sea shear zone. In: B. Zanettin (ed.) *Geodynamic Evolution of the Afro-Arabian Rift System,*. 1979, **47**, 666. Rome: *Atti Dei Convegni Lincei.*

Barberi, F., Civetta, L., and Varet, J. (1980) Sr isotope composition of Afar volcanics and its implication for mantle evolution. *Earth Planet. Sci. Lett.* **50**, 247.

Barnard, P.C., Thompson, S., Bastow, M.A., Ducreux, C., and Mathurin, G. (1992) Thermal maturity development and source-rock occurrence in the Red Sea and Gulf of Aden. *J. Petrol. Geol.* **15**, 173.

Barrat, J.-A., Jahn, B.M., Joron, J.-L., Auvray, B., and Hamdi, H. (1990) Mantle heterogeneity in northeastern Africa: evidence from Nd isotopic compositions and hygromagmaphile element geochemistry of basaltic rocks from the Gulf of Tadjoura and southern Red Sea regions. *Earth Planet. Sci. Lett.* **101**, 233.

Bartov, Y. (1980) Sinistral movement along the Gulf of Aqaba–its age and relation to opening of the Red Sea. *Nature* **285**, 220.

Bayer, H.-J., Hotzl, H., Jado, A.R., Roscher, B., and Voggenreiter, W. (1988) Sedimentary and structural evolution of the northwest Arabian Red Sea margin. *Tectonophysics* **153**, 137.

Bayer, H.-J., El-Isa, Z., Hotzl, H., Prodehl, C., and Saffarini, G. (1989) Large tectonic and lithospheric structures of the Red Sea region. *J. Afri. Earth Sci.* **8**, 565.

Ben-Avraham, Z. (1985) Structural framework of the Gulf of Elat (Aqaba), Northern Red Sea. *J. Geophys. Res.* **90**, 703.

Ben-Avraham, Z. (1987) Sedimentary basins within the Dead Sea and other rift zones. *Tectonophysics* **141**, 1.

Ben-Avraham, Z., and Von Herzen, R.P. (1987) Heat flow and continental breakup: the Gulf of Elat (Aqaba). *J. Geophys. Res.* **92**, 1407.

Ben-Avraham, Z., Hänel, R., and Villinger, H. (1978) Heat flow through the Dead Sea rift. *Mar. Geol.* **28**, 253.

Ben-Avraham, Z., Almagor, G., and Garfunkel, Z. (1979a) Sediments and structure of the Gulf of Elat (Aqaba), Northern Red Sea. *Sediment. Geol.* **23**, 239.

Ben-Avraham, Z., Garfunkel, Z., Almagor, G., and Hall, J.K. (1979b) Continental breakup by a leaky transform: the Gulf of Elat (Aqaba). *Science* **206**, 214.

Bender, F. (1975) Jordan, Geology of the Arabian Peninsula. *U.S. Geological Survey Prof. Paper* no. **560-D**.

Berckhemer, H., Baier, B., Bartelsen, H., Behle, A., Burkhardt, H., Gebrande, H., Makris, J., Menzel, H., Miller, H., and Vees, R. (1975) Deep seismic soundings in the Afar region and on the highland of Ethiopia. In: *Afar Depression of Ethiopia*. Stuttgart: Schweizerbart. p.89.

Berhe, S.M. (1986) Geologic and geochronologic constraints on the evolution of the Red Sea–Gulf of Aden and Afar depression. *J. Afri. Earth Sci.* **5**, 101.

Berthon, J.L., Burollet, P.F., and Legrand, P. (1987) *Genese et evolution des bassins sedimentaires*. Paris: TOTAL Compagnie Francaise des Petroles.

Betton, P.J., and Civetta, L. (1984) Strontium and neodymium isotopic evidence for the heterogenous nature and development of the mantle beneath Afar (Ethiopia). *Earth Planet. Sci. Lett.* **71**, 59.

Beydoun, Z.R. (1988) *The middle east: Regional geology and petroleum resources.* Beaconsfield, Bucks: Scientific Press.

Beydoun, Z.R. (1989) Hydrocarbon prospects of the Red Sea–Gulf of Aden. *J. Petrol. Geol.* **12**, 125.

Beydoun, Z.R., and Sikander, A.H. (1992a) The Red Sea–Gulf of Aden: re-assessment of hydrocarbon potential. *J. Petrol. Geol.* **15**, 245.

Beydoun, Z.R., and Sikander, A.H. (1992b) The Red Sea–Gulf of Aden: re-assessment of hydrocarbon potential. *Mar. Petrol. Geol.* **9**, 474.

Beyth, M. (1991) "Smooth" and "rough" propagation of spreading: southern Red Sea–Afar depression. *J. Afr. Earth Sci.* **13**, 157.

Bignell, R.D., Cronan, D.S., and Tooms, J.S. (1976) Red sea metalliferous brine precipi- tates. *Geol. Soc. Can. Spec. Paper* **14**, 147.

Bischoff, J.L. (1969) Red Sea geothermal brine deposits: their mineralogy, chemistry, and genesis.In E.T. Degens and D.A. Ross (eds) *Hot Brines and Recent Heavy Metal Deposits in the Red Sea*. New York: Springer-Verlag, p. 368.

Black, R., Morton, W.H., and Varet, J. (1972) New data on Afar tectonics. *Nat. Phys. Sci.* **240**, 170.

Blank, H.R.J. (1977). Areomagnetic and geologic study of Tertiary dikes and related structures on the Arabian margin of the Red Sea. In: *Red Sea Research 1970– 1975* (22) G1. Jiddah: *Saudi Arabian Directorate General of Mineral Resources Bull.* **22**, G1-G18.

Bohannon, R.G. (1986a) Tectonic configuration of the western Arabian continental mar- gin, southern Red Sea. *Tectonics* **5**, 477.

Bohannon, R.G. (1986b) How much divergence has occurred between Africa and Arabia as a result of the opening of the Red Sea? *Geology* **14**, 510.

Bohannon, R.G., and Eittreim, S.L. (1991) Tectonic development of passive margins of the southern and central Red Sea with a comparison to Wilkes Land, Antarctica. *Tectonophysics* **198**, 129.

Bohannon, R.G., Naeser, C.W., Schmidt, D.L., and Zimmerman, R.A. (1989) The timing of uplift, volcanism, and rifting peripheral to the Red Sea: a case for passive rifting? *J. Geophys. Res.* **94**, 1683.

Bonatti, E. (1985) Punctiform initiation of seafloor spreading in the Red Sea during transition from a continental to an oceanic rift. *Nature* **316**, 33.

Bonatti, E. (1988) Zabargad Island and the Red Sea rift (*Collected papers*). *Tectonophysics* **150**, 1–260

Bonatti, E., Hamlyn, P.R., and Ottonello, G. (1981) The upper mantle beneath a young oceanic rift: peridotites from the island of Zabargad (Red Sea). *Geology* **9**, 474.

Bonatti, E., Clocchiatti, R., Colantini, P., Gelmini, R., Marrinelli, G., Ottonello, G., Santacroce, R., Tahin, M., Abdel-Meguid, A.A., Assaf, H.S., and El Tabir, M.A. (1983) Zabargad (St. John's) Island: an uplifted fragment of sub-Red Sea lithosphere. *J. Geol. Soc. Lond.* **140**, 677.

Bonatti, E., Colantoni, P., Vedova, D. and Tavani, M. (1984) Geology of the Red Sea transitional zone (22°N–25°N). *Oceanogr. Acta* **7**, 385.

Bonatti, E., Ottonello, G., and Hamlyn, P.R. (1986) Peridotites from the island of Zabargad (St. John), Red Sea: petrology and geochemistry. *J. Geophys. Res.* **91**, 599.

Bosworth, W. (1985) Geometry of propagating continental rifts. *Nature* **316**, 625.

Bott, W.F., Smith, B.A., Oakes, G., Sikander, A.H., and Ibraham, A.I. (1992) The tectonic framework and regional hydrocarbon prospectivity of the Gulf of Aden. *J. Petrol. Geol.* **15**, 211.

Boudier, F., Nicolas, A., Ji, S., Kienst, J.R., and Mevel, C. (1988) The gneiss of Zabargad Island: deep crust of a rift. *Tectonophysics* **150**, 209.

Bowen, R. (1984a) Meeting reviews African geology. *Geotimes* **29**(9), 13.

Bowen, R. (1984b) Geological problems of north-east Africa. *Nature* **311**, 108.

Bowen, R., and Jux, U. (1987) *Afro-Arabian Geology: A Kinematic View.* New York: Chapman and Hall.

Braithwaite, C.J.R. (1987) Geology and paleogeography of the Red Sea region. In: A J. Edwards and S.M. Mead (eds.) *Key environments: Red Sea.* New York: Pergamon Press, p. 22.

Brinckmann, J., and Kursten, M. (1969) *Geological sketchmap of the Danakil depression* (1:250,000). Hanover: Bundesanstaldt fur Bodenforschung.

Brooks, C.K., and Nielsen, T.F.D. (1982) The E. Greenland continental margin: a transition between oceanic and continental magmatism. *J. Geol. Soc. Lond.* **139**, 265.

Brown, G.F. (1970) Eastern margin of the Red Sea and the coastal structures in Saudi Arabia. *Phil. Trans. R. Soc. Ser. A* **267**, 75.

Brown, G.F. (1972) *Tectonic Map of the Arabian Peninsula:Map AP-2, scale 1:4,000,000.* Jiddah: Saudi Arabian Directorate General of Mineral Resources.

Brown, G.F., Schmidt, D.L., and Huffman, A.C.J. (1989) Shield area of western Saudi Arabia, Geology of the Arabian Peninsula. *U.S. Geological Survey Prof.* Paper. no. **560-A**.

Brueckner, H.K., Zindler, A., Seyler, M., and Bonatti, E. (1988) Zabargad and the isotopic evolution of the sub-Red Sea mantle and crust. *Tectonophysics* **150**, 163.

Bruneau, L., Jerlov, N.G., and Koczy, F.F. (1953) Physical and chemical methods. *Reports of the Swedish Deep-Sea Expedition.* Goteburg: Vitterhets-Samhalle, pp.101-102, appendix I–LV.

Buck, W.R. (1986) Small-scale convection induced by passive rifting: the cause for uplift of rift shoulders, *Earth Planet. Sci. Lett.* **77**, 362.

Buck, W.R., Martinez, M.S., Steckler, M., and Cochron, S.R. (1988) Thermal consequences of lithospheric extension: pure and simple. *Tectonics* **7**, 213.

Bunter, M.A.G., and Abdel Magrid, A.E.M. (1989a) The Sudanese Red Sea: 1. New developments in stratigraphy and petroleum-geological evolution. *J. Petrol. Geol.* **12**, 145.

Bunter, M.A.G., and Abdel Magrid, A.E.M. (1989b) The Sudanese Red Sea: 2. New developments in the petroleum geochemistry. *J. Petrol. Geol.* **12**, 167.

Camp, V.E. (1984) Island arcs and their role in the evolution of the western Arabian Shield. *Bull.Geol. Soc. Amer.* **95**, 913.

Camp, V.E., and Roobol, M.J. (1989) The Arabian continental alkali basalt province: Part I: Evolution of Harrat Rahat, Kingdom of Saudi Arabia. *Bull. Geol. Soc. Amer.* **101**, 71.

Camp, V.E., Hooper, P.R., Roobol, M.J., and White, D.L. (1987) The Medina eruption, Saudi Arabia: magma mixing and simultaneous extrusion of three basaltic chemical types. *Bull. Volcanol.* **49**, 489.

Camp, V.E., Roobol, M.J., and Hooper, P.R. (1991) The Arabian continental alkali basalt province: Part II. Evolution of Harrats Khaybar, Ithnayn, and Kura, Kingdom of Saudi Arabia. *Bull. Geol. Soc. Amer.* **103**, 363.

Camp, V.E., Roobol, M.J., and Hooper, P.R. (1992) The Arabian continental alkali basalt province: Part III. Evolution of Harrat Kishb, Kingdom of Saudi Arabia. *Bull. Geol. Soc. Amer.* **104**, 379.

Capaldi, G., Manetti, P., and Piccardo, G.B. (1983) Preliminary investigations on volcanism of the Sadah region (Yemen Arabic Republic). *Bull. Volcanol.* **46**, 413.

Capaldi, G., Chiesa, S., Manetti, P., Orsi, G. and Poli, G. (1987a) Tertiary anorogenic granites of the western border of the Yemen plateau. *Lithos.* **20**, 433.

Capaldi, G., Manetti, P., Piccardo, G.B., and Poli, G. (1987b) Nature and geodynamic significance of the Miocene dyke swarm in the North Yemen. *N. Jb. Miner. Abh.* **156**, 207.

Carella, R. and Scarpa, N. (1962) *Geologic results of exploration in Sudan by A.G.I.P. Mineraria Ltd.* In: *Fourth Arabian Petroleum Congress, Beirut.* Milan: Donato.

Carlson, R.W. (1984) Isotopic constraints on Columbia River flood basalt genesis and the nature of the subcontinental mantle. *Geochem. Cosmochim. Acta* **48**, 2357.

Chase, R.L. (1969) Basalt from the axial trough of the Red Sea. In: E.T. Degens and D.A Ross (eds) *Hot Brines and Recent Heavy Metal Deposits in the Red Sea.* New York: Springer-Verlag, p. 122.

Chiesa, S., La Volpe, L., Lirer, L., and Orsi, G. (1983) Geological and structural outline of Yemen plateau; Yemen Arab Republic. *N.Jb. Geol. Paleont. Mh.* **11**, 641.

Chiesa, S., Civetta, L., De Fino, M., La Volpe, L., and Orsi, G. (1989) The Yemen Trap Series: genesis and evolution of a continental flood basalt province. *J. Volcanol. Geotherm. Res.* **36**, 337.

Cita, M.B., and Wright, R., (eds) (1979) Geodynamic and biodynamic effects of the Messinian salinity crisis in the Mediterranean. *Paleog. Paleoecol.* **29**.

Civetta, L., La Volpe, L., and Lirer, L. (1978) K-Ar ages of the Yemen plateau. *J. Volcanol. Geotherm. Res.* **4**, 307.

Civetta, L., De Fino, M., La Volpe, L., and Lirer, L. (1979) Recent volcanism of North Yemen: structural and genetic implications. In: *Geodynamic Evolution of the Afro-Arabian rift system. Atti Dei Convegni Lincei.* **47**, 617.

Clifford, T.N. (1970) The structural framework of Africa. In: T.N. Clifford and I.G. Grass (eds) *African Magmatism and tectonics.* Edinburgh: Oliver and Boyd, p. 1.

Clin, M. (1991) Evolution of eastern Afar and the Gulf of Tadjura. *Tectonophysics* **198**, 355.

Clin, M., Pouchan, P. (1970) *Carte Geologique du Territore Française des Afars et des Issas, 1:200,000.* Djibouti: Centre d'Etudes Geologique et de Development du T.F.A.I.

Cloos, H. (1930) Zur experimentellen Tektonik. *Naturwissenschaften* **34**, 741.

CNR (1975) *Geology of central and southern Afar, map at scale 1:500,000.* Paris: CNR.

CNR CNS Afar team (1973) Geology of northern Afar (Ethiopia). *Rev. Geog. Phys. Geol. Dyn.* **15**, 443.

Cochran, J.R. (1981) The Gulf of Aden: structure and evolution of a young ocean basin and continental margin. *J. Geophys. Res.* **86**, 263.

Cochran, J.R. (1983) A model for the development of the Red Sea. *Amer. Assoc. Petrol. Geol. Bull.* **67**, 41.

Cochran, J.R., and Martinez, F. (1988) Evidence from the northern Red Sea on the transition from continental rifting to seafloor spreading. *Tectonophysics* **153**, 25.

Cochran, J.R., Martinez, F., Steckler, M.S., and Hobart, M.A. (1986) Conrad Deep: a new northern Red Sea deep. Origin and implications for continental rifting. *Earth Planet. Sci. Lett.* **78**, 18.

Cochran, J.R., Gaulier, J.-M., and Le Pichon, X. (1991) Crustal structure and the mechanism of extension in the northern Red Sea: constraints from gravity anomalies. *Tectonics* **10**, 1018.

Coleman, R.G. (1974) Geological background of the Red Sea. *Initial Reports of the Deep Sea Drilling Project* **23**, 813.

Coleman, R.G. (1984a) The Red Sea: a small ocean basin formed by continental extension and sea-floor spreading. In: N.A. Bogdanov (ed.) *Proceedings of the 27th International Geological Congress 27th August 1984, Moscow* **23**, 93. Utrecht: VNU Science Press.

Coleman, R.G. (1984b) The Tihama Asir igneous complex, a passive margin ophiolite. *Proceedings of the 27th International Geological Congress 27th August 1984, Moscow* **9**, 221. Utrecht: VNU Science Press.

Coleman, R.G., and McGuire, A.V. (1988) Magma systems related to the Red Sea opening. *Tectonophysics* **150**, 77.

Coleman, R.G., Tatsumoto, M., Coles, D.G., Hedge, C.E., and Mays, R.E. (1974) Red Sea basalts. *EOS Trans. Amer. Geophys. Union* **54**, 1001.

Coleman, R.G., Fleck, R.J., Hedge, C.E., and Ghent, E.D. (1977) The volcanic rocks of southwest Saudi Arabia and the opening of the Red Sea. *Saudi Arabia Min. Res. Bull.* **22**, D1–D30

Coleman, R.G., Gregory, R.T., Fleck, R.J., Hedge, C.E., Donato, M.M. (1979) The Miocene Tihama Asir ophiolite and its bearing on the opening of the Red Sea. *King Abdulaziz University Institute of Applied Geology, I.A.G. Bull.* **1**, 173.

Coleman, R.G., Gregory, R.T., and Brown, G.F. (1983) Cenozoic Volcanic Rocks of Saudi Arabia. *U.S. Geological Survey Open-File Report,* OF-03-93,

Coleman, R.G., DeBari, S. and Peterman, Z. (1992). A-type granite and the Red Sea opening. *Tectonophysics* **204**, 27.

Colleta, B., Le Quellec, P., Letouzey, J., and Moretti, I. (1988) Longitudinal evolution of the Suez rift structure (Egypt). *Tectonophysics* **153**, 221.

Courtillot, V.E. (1980) Opening of the Gulf of Aden and Afar by progressive tearing. *Phys. Earth Planet. Interiors.* **21**, 343.

Courtillot, V.E., Armijo, R., and Tapponnier, P. (1987) Kinematics of the Sinai triple junction and a two-phase model of Arabia–Africa rifting. *Geol. Soc. Lond. Spec. Pub.* **28**, 559.

Cox, K.G., Gass, I.G., and Mallick, D.I.J., (1969) The evolution of the volcanoes of Aden and Little Aden, South Arabia. *J. Geol. Soc. Lond.* **124**, 283.

Cox, K.G., Gass, I.G., and Mallick, D.I.J., (1970) The peralkaline volcanic suite of Aden and Little Aden, South Arabia. *J. Petrol.* **11**, 433.

Craig, H. (1969) Geochemistry and origin of the Red Sea brines. In Degans and Ross (eds) *Hot Brines and Recent Heavy Metal Deposits in the Red Sea.* New York: Springer-Verlag p. 208.

Crossley, R., Watkins, C., Raven, M., Cripps, D., Carnell, A., and Williams, D. (1992) The sedimentary evolution of the Red Sea and Gulf of Aden. *J. Petrol. Geol.* **15**, 157.

Davidson, A., and Rex, D.C. (1980) Age of volcanism and rifting in southwest Ethiopia. *Nature* **283**, 657–658.

Davies, D., and Tramontini, C. (1969) A seismic refraction survey in the Red Sea. *Geophys. J. R. Astr. Soc.* **17**, 225.

Davies, D., and Tramontini, C. (1970) The deep structure of the Red Sea. *Phil. Trans. Roy. Soc. Ser. A* **267**, 181.

Degens, E.T., and Ross, D.A. (eds) (1969) *Hot Brines and Recent Heavy Metal Deposits in the Red Sea.* New York: Springer-Verlag.

Delaney, F.M. (1954) Recent Contributions to the Anglo-Egyptian Sudan. *19th Int. Geol. Congress, Algiers* **20**, 11.

Dercourt, J., Zonenshain, L.P., Ricou, L.E., Kazmin, V., Le Pichon, X., Knipper, A.L., Grandjacquet, C., Sbortshikov, J.M., Geyssant, J., Lepurier, C., Pechersky, D.H., Boulin, J., Sibuet, J.C., Savostin, L.A., Sorokhtin, O., Westphal, M., Bazhenov, M., Lauer, L., and Biju-Duval, B. (1986) Geologic evolution of the Tethys from the Atlantic to the Pamirs since Lias. *Tectonophysics* **123**, 241.

Dickinson, D.R., Dodson, M.H., Gass, I.G., and Rex, D.C. (1969) Correlation of initial $^{87}Sr/^{86}Sr$ with Rb/Sr in some late Tertiary volcanic rocks of south Arabia. *Earth Planet. Sci. Lett.* **6**, 84.

Dixon, T.H., Stern, R.J., and Hussein, I.M. (1987) Control of Red Sea rift geometry by Precambrian structures. *Tectonics* **6**, 551.

Dixon, T.H., Ivins, E.R., and Franklin, B.J. (1989) Topographic and volcanic asymmetry around the Red Sea: constraints on rift models. *Tectonics* **8**, 1193.

Doughty, C.M. (1979) *Travels in Arabia Deserta.* New York: Dover.

Drake, C.L., and Girdler, R.W. (1964) A geophysical study of the Red Sea. *Geophys. J. R. Astr. Soc.* **8**, 473.

Du Bray, E.A., Stosser, D., and McKee, E.H. (1991) Age and petrology of the Tertiary As Sirat volcanic field, southwestern Saudi Arabia. *Tectonophysics* **198**, 155.

Duffield, W.A., McKee, E.H., El Salem, F., and Teimeh, M. (1988) K-Ar ages, chemical composition and geothermal significance of Cenozoic basalt near the Jordan rift. *Geothermics* **17**, 635.

Durozoy, G. (1972) Hydrogeologic des basalts du Harrat Rahat. *France, Bur. de Rec. Geol. et Min. Bull.* **Ser. 2, sect. 3**(2), 37.

Edmond, J.M., Von Damm, K.L., McDuff, R.E., and Measures, C.I. (1982) Chemistry of hot springs on the East Pacific rise and their effluent dispersal. *Nature* **297**, 187.

Edwards, A.J., and Herad, S.M. (1987) *Key Environments: Red Sea.* Oxford: Pergamon Press.

Egloff, F., Rihm, R., Makris, J., Izzeldin, Y.A., Bobsien, M., Meier, K., Junge, P., Norman, T., and Warsi, W. (1991) Contrasting structural styles of the eastern and western margins of the southern Red Sea: the 1988 SONNE experiment. *Tectonophysics* **198**, 329.

El-Isa, Z.H., Makris, J., and Prodehl, C. (1986) A deep seismic sounding experiment in Jordan. *Dirasat* **13**, 271.

El-Isa, Z., Mechie, J., and Prodehl, C. (1987a) Shear velocity structure of Jordan from explosion seismic data. *Geophys. J. R. Astr. Soc.* **90**, 265.

El-Isa, Z., Mechie, J., Prodehl, C., Makris, J., and Rihm, R. (1987b) A crustal structure study of Jordan derived from seismic refraction data. *Tectonphysics* **138**, 235.

Emery, K.O., Hunt, J.M., and Hays, E.E. (1969) Summary of hot brines and heavy metal deposits in the Red Sea. In: E.T. Degens and D.A. Ross (eds) *Hot Brines and Recent Heavy Metal Deposits in the Red Sea.* New York: Springer-Verlag, p. 557.

Esperanca, S., and Garfunkel, Z. (1986) Ultramafic xenoliths from the Mt. Carmel area (Karem Maharal volcano), Israel. *Lithosphere* **19**, 43.

Evans, A.L. (1988) Neogene tectonic and stratigraphic events in the Gulf of Suez rift area, Egypt. *Tectonophysics* **153**, 235.

Eyal, M., Bartov, Y., Shimron, A.E., and Bentor, Y.K. (1980) *Geological Map of the Sinai, scale 1:500,000.* Geological Survey of Israel.

Eyal, M., Eyal, Y., Bartov, Y., and Steinitz, G. (1981) The tectonic development of the western margin of the Gulf of Elat (Aqaba) rift. *Tectonophysics* **80**, 39.

Falcon, N.L., Gass, I.G., Girdler, R.W., and Laughton, A.S. (eds) (1969) A discussion on the structure and evolution of the Red Sea and the nature of the Red Sea, Gulf of Aden and Ethiopian rift junction. (*Collected Papers*) *Phil. Trans. R. Soc. Ser A* **267**, 1–417

Filjak, R., Glumick, N., and Zagorac, Z. (1959) Oil possibilities of the Red Sea region in Ethiopia. *Zagrevb., Naftaplin.*104.

Fleck, R.J., and Hadley, D.G. (1982) Ages and strontium initial ratios of plutonic rocks in a transect of the Arabian shield. *Saudi Arabian Deputy Ministry, Mineral Resources, Open File Report. USGS,* **OF-03-38**, 43.

Francheteau, J., Needham, H.D., Juteau, T., Seguret, M., Ballard, R.D., Fox, P.J., Normark, W., Carranza, A., Cordoba, G. J., Rangin, C., Bouggault, B., Cambon, P., and Hekinian, R. (1979) Massive deep-sea sulphide ore deposits discovered on the East Pacific Rise. *Nature* **277**, 523.

Frazier, S.B. (1970) Adjacent structures of Ethiopia: that portion of the Red Sea coast including Dahlak Kebir Island and the Gulf of Zula. *Phil. Trans. R. Soc. Ser. A* **267**, 131.

Freund, R. (1965) A model of the structural development of Israel and adjacent areas since the upper Cretaceous time. *Geol. Mag.* **189,** 205

Gangi, A.F. (1989) World rift systems (A symposium). *Tectonophysics* **197**, 391.

Garfunkel, Z. (1981) Internal structure of the Dead Sea leaky transform (rift) in relation to plate tectonics. *Tectonophysics* **80**, 81.

Garfunkel, Z. (1988) Relation between continental rifting and uplifting: evidence from the Suez rift and northern Red Sea. *Tectonophysics* **150**, 33.

Garfunkel, Z., and Bartov, Y. (1977) The tectonics of the Suez rift. *Geol. Surv. Israel Bull.* **71**.

Garfunkel, Z., Bartov, J., Eyal, Y., and Steinitz, G. (1976) Raham conglomerate–new evidence for Neogene tectonism in the southern part of the Dead Sea rift. *Geol. Mag.* **111**, 55.

Garfunkel, Z., Ginzburg, A., and Searle, R.C. (1987) Fault pattern and mechanism of crustal spreading along the axis of the Red Sea from side scan [Gloria] data. *Ann. Geophys.* **5B**, 187.

Gass, I.G. (1970) The Evolution of Volcanism in the Junction of the Red Sea, Gulf of Aden and Ethiopean rifts. *Phil. Trans. R. Soc. Ser. A* **267**, 369.

Gass, I.G. (1977) The age and extent of the Red Sea oceanic crust. *Nature* **265**, 722.

Gass, I.G. (1980) *Crustal and Mantle Processes–Red Sea Case Study. An Embryonic Ocean Basin.* Milton Keynes: Open University

Gass, I.G., Mallick, D.I.J., and Cox, K.G. (1973) Volcanic islands of the Red Sea. *J. Geol. Soc. Lond.* **129**, 275.

Gaulier, J.-M., Le Pichon, X., Lyberis, N., Avedik, F., Gelic, L., Moretti, I., Deschampes, A., and Hafez, S. (1988) Seismic study of the crust of the northern Red Sea and Gulf of Suez. *Tectonophysics* **153**, 55.

Gettings, M.E. (1977) Delineation of the continental margin in the southern Red Sea from new gravity evidence. *Red Sea Research 1970–1975.* Jiddah: Saudi Arabian Directorate General of Mineral Resources Bull. K1-K11.

Gettings, M.E., Blank, H.R.J., and Mooney, W.D. (1986) Crustal structure of southwestern Saudi Arabia. *J. Geophys. Res.* **91**, 6491.

Geukens, F. (1966) Yemen, Geology of the Arabian Peninsula. *U.S. Geological Survey Prof. Paper.* no. **560-B**, 23 C.

Ghent, E.D., Coleman, R.G., and Hadley, D.G. (1980) Ultramafic inclusions and host alkali olivine basalts of the southern coastal plain of the Red Sea, Saudi Arabia. *Amer. J. Sci.* **280-A**, 499.

Gillmann, M. (1968) Primary results of a geological and geophysical reconnaissance of the Jizan coastal plain in Saudi Arabia. In: *Proceedings of the 2nd AIME Regional Technical Symposium, Dhahran,* 189.

Ginzburg, A., Makris, A., Fuchs, J., Prodehl, C., Kaminski, W., and Amitai, U. (1979a) A seismic study of the crust and upper mantle of the Jordan–Dead Sea rift and their transition toward the Mediterranean Sea. *J. Geophys. Res.* **84**, 1569.

Ginzburg, A., Makris, J., Fuchs, K., Perthoner, B., and Prodehl, C. (1979b) Detailed structure of the crust and upper mantle along the Jordan–Dead Sea Rift. *J. Geophys. Res.* **84**, 5605.

Girdler, R.W. (1970a) An areomagnetic survey of the junction of the Red Sea, Gulf of Aden and Ethiopian rifts (preliminary report). *Phil. Trans. R. Soc. Ser. A* **267**, 359.

Girdler, R.W. (1970b) A review of Red Sea heat flow. *Phil. Trans. R. Soc. Ser. A* **267**, 191.

Girdler, R.W. (1983) The evolution of the Gulf of Aden and Red Sea in space and time. In: M.V. Angel (ed.) *Marine Science of the North-West Indian Ocean and Adjacent Waters.* Egypt: Pergamon Press, p. 747.

Girdler, R.W. (1985) Problems concerning the evolution of the oceanic lithosphere in the northern Red Sea. *Tectonophysics* **116**, 109.

Girdler, R.W. (1991) The case for ocean crust beneath the Red Sea. *Tectonophysics* **198** 275.

Girdler, R.W., and Darracott, B.W. (1972) African poles of rotation, comments on earth sciences. *Geophysics* **2**, 131.

Girdler, R.W., and Evans, T.R. (1977) Red Sea heat flow. *Geophys, J. R. Astron. Soc.* **51**, 245.

Girdler, R.W., and Hall, S.A. (1972) An aeromagnetic survey of the Afar triangle of Ethiopia. *Tectonophysics* **15**, 53.

Girdler, R.W., and Harrison, J.C. (1957) Submarine gravity measurements in the Atlantic Ocean, Indian Ocean, Red Sea, and Mediterranean Sea. *Proc. R. Soc. Ser. A* **239**, 202.

Girdler, R.W., and Styles, P. (1974) Two-stage Red Sea floor spreading. *Nature* **247**, 1.

Girdler, R.W. and Styles, P. (1976) Opening of the Red Sea with two poles of rotation–some comments. *Earth Planet. Sci. Lett.* **33**, 169.

Girdler, R.W., and Styles, P. (1978) Sea floor spreading in the western Gulf of Aden. *Nature* **271**, 615.

Girdler, R.W., and Underwood, M. (1985) The evolution of early oceanic lithosphere in the southern Red Sea. *Tectonophysics* **116**, 95.

Girdler, R.W., Whitmarsh, R.B. (1974) Miocene evaporites in Red Sea cores: their relevance to the problem of the width and age of oceanic crust beneath the Red Sea. *Int. Reports Deep Sea Drilling Project* **23**, 913-922.

Gouin, P. (1970) Seismic and gravity data from Afar in relation to surrounding areas. *Phil. Trans. Roy. Soc. Ser. A* **267**, 339.

Greenwood, J.E.G.W., and Bleackley, D. (1967) Aden Protectorate, Geology of the Arabian Peninsula. *U.S. Geological Survey Prof. Paper.* no. **560-C**, 96 C.

Greenwood, W.R. (1973) The Ha'il arch – a key to the deformation of the Arabian Shield during the evolution of the Red Sea rift. *Saudi Arabian Dir. Gen. Miner. Resour. Bull.* **7** (5).

Greenwood, W.R., Hadley, D.G., Anderson, R.E., Fleck, R.J., and Schmidt, D.L. (1976) Late Proterozoic cratonization in southwestern Saudi Arabia. *Phil. Trans. R. Soc. Ser. A.* **280**, 517.

Greiling, R.O., El Ramly, M.F., El Akhal, H., and Stern, R.J. (1988) Tectonic evolution of the northwestern Red Sea margin as related to basement structure. *Tectonophysics* **153**, 179.

Grolier, M.J., and Overstreet, W.C. (1978) *Geologic Map of Yemen Arab Republic (San'a),* U.S. Geological. Survey, Misc. Inv. Series, Map 1-1143-B, Scale 1:500,000. U.S. Geological Survey.

Guennoc, P., Pautot, G., and Coutelle, A. (1988) Surficial structures of the northern Red Sea axial valley from 23° to 28°N: time and space evolution of neo-oceanic structures. *Tectonophysics* **153**, 1.

Guennoc, P., Pautot, G., Le Qentrec, M.F., and Coutelle, A. (1990) Structure of an early oceanic rift in the northern Red Sea. *Oceanol. Acta.* **13**, 145.

Guney, M., Al-Marhoun, M.A., and Nawab, Z.A. (1988) Metalliferous submarine sediments of the Atlantis-II-Deep, Red Sea. *Can. Inst. Mining Bull.* **81**, 33.

Hadley, D.G., and Schmidt, D.L. (1980) Sedimentary rocks and basins of the Arabian Shield and their evolution. *Inst. Applied Geology, King Abdulaziz Univ. Bull.* **4**, 25.

Hadley, D.G., Schimdt, D.L., and Coleman, R.G. (1982) Summary of Tertiary investigations in western Saudi Arabia, current work by the U.S. Geological Survey, and recommended future studies. *U.S. Geological Survey Open File Report.* no. **USGS-OF-03-5**.

Haenel, R. (1972) Heat flow measurements in the Red Sea and Gulf of Aden. *Z. Geophys.* **38**, 1035.

Haitham, F.M.S., and Nani, A.S.O. (1990) The Gulf of Aden rift: hydrocarbon potential of the Arabian sector. *J. Petrol. Geol.* **13**, 211.

Hall, S.A. (1989) Magnetic evidence for the nature of the crust beneath the southern Red Sea. *J. Geophys. Res.* **94**, 12, 267.

Hall, S.A., Andreason, G.E., and Girdler, R.W. (1977) Total intensity magnetic anomaly map of the Red Sea and adjacent coastal areas, a description and preliminary interpretation. *Saudi Arabia Dir. Gen. Mineral resources Bulletin (Red Sea research 1970-1975)* **22**, F1-F15.

Hart, W.K., Wolde-Gabriel, G., Walter, R.C., and Mertzman, S.A. (1989) Basaltic volcanism in Ethiopia: constraints on continental rifting and mantle interactions. *J. Geophys. Res.* **94**, 7731.

Hartmann, M. (1980) Atlantis II Deep geothermal brine system. Hydrographic situation in 1977 and changes since 1965. *Deep Sea Res.* **27A**, 161.

Haymon, R.M. (1989) Hydrothermal processes and products on the Galapagos rift and East Pacific rise. In: *The Geology of North America.* Boulder Co: Geological Society of America, p. 125.

Haymon, R.M., and Macdonald, K.C. (1985) The geology of deep-sea hot springs. S*cien. Amer.* **73**, 441.

Head, S.M. (1987) Corals and coral reefs of the Red Sea. In: *Key Environments: Red Sea.* New York: Pergamon Press, p.128.

Healy, J.H., Mooney, W.D., Blank, H.R.J, Gettings, M.E., Kholer, W.M., Lamson, R.J., and Leone, L.E. (1982) S*audi Arabian Seismic Deep-Refraction Profile:Final Project Report. U.S. Geological Survey Open-File report.* no.6 **USGS-OF-02-37**.

Hegner, E., and Pallister, J.S. (1989) Pb, Sr, and Nd isotopic characteristics of Tertiary Red Sea rift volcanics from Central Saudi Arabian Coastal Plain. *J. Geophys. Res.* **94**, 7749.

Helms, S.W. (1981) *Jawa, Lost City of the Black Desert.* Ithaca, NY: Cornell University Press.

Helmy, H.M. (1990) Southern Gulf of Suez, Egypt: structural geology of the B-trend oil fields. In *Classic Petroleum Provinces.* London: Alden Press, p.353.

Hempton, M.R. (1987) Constraints on Arabian plate motion and extensional history of the Red Sea. *Tectonics* **6**, 687.

Henry, C., and Chorowicz, J. (1986) *Region du Sinai, Carte geologique et geomorphologue teleanalytique au 1:500,000 d'apres les images Landsat.* Paris: Technip.

Heybroek, F. (1965) The Red Sea evaporite basin. In: *Salt Basins around Africa.* London: Institute of Petroleum, p.17.

Hilpert, L.S. (1977) Red Sea research 1970–1975. *Saudi Arabian Directorate General of Mineral Resources Bulletin.* **22** (A1-A2) (Collected Papers).

Hsu, K., and Bernoulli, D. (1978) Genesis of the Tethys and the Mediterranean. *Int. Rept. Deep Sea Drilling Project* **42**, 943.

Hughes, G.W., and Beydoun, Z.R. (1992) The Red Sea–Gulf of Aden: biostratigraphy and paleoenvironments. *J. Petrol. Geol.* **15**, 135.

Hughes, G.W., Varol, O., and Beydoun, Z.R. (1991) Evidence for Middle Oligocene rifting of the Gulf of Aden and for Late Oligocene rifting of the southern Red Sea. *Mar. Petrol. Geol.* **8**, 354.

Hutchinson, R.W., and Engels, G.G. (1970) Tectonic significance of regional geology and evaporite lithofacies in northeastern Ethiopia. *Phil. Trans. R. Soc. Ser. A* **267**, 313.

Izzeldin, Y.A. (1987) Seismic, gravity and magnetic surveys in the central part of the Red Sea: their interpretation and implications for the structure and evolution of the Red Sea. *Tectonophysics* **143**, 269.

Jobert, G. (1980) Resultats Geophysiques dans l'Afar et les regions voisimes, in colloque Rift D'Asal. *Soc. Geol. France Bull.* **22**, 1003.

Joffe, S., and Garfunkel, Z. (1987) Plate kinematics of the circum Red Sea – a re-evaluation. *Tectonophysics* **141**, 5.

Jones, P.W. (1976) Age of the lower flood basalts of the Ethiopian Plateau. *Nature* **261**, 567.

Juteau, T., Eissen, J.P., Monin, A.S., Zonenshain, L.P., Sorokhtin, O.G., Matveenkov, V.V., and Aluymukhamedov, A.I. (1983) Structure and petrology of the Red Sea axial rift at 18° North: results of the Soviet diving program with submersible (1980). *Bull. Centres Rech. Explor. Prod. Elf-Aquitaine* **7**, 217.

Karbe, L. (1987) Hot brines and deep sea environment. In: *Key Environments: Red Sea.* New York: Pergamon Press, p.90.

Karpoff, R. (1955) Esquisse geologique de l'Arabie Seoudite. *Geol. Soc. France* **7**(6 ser.), 672.

Kazmin, V. (1973) *Geological map of Ethiopia, scale 1:2,000,000.* Addis Ababa: Ministry of Mines, Geological Survey of Ethiopia.

Kazmin, V., Siefe, M.B., Nicoletti, M., and Petucciani, C. (1980) Evolution of the northern part of the Ethiopian rift. In: Geodynamic Evolution of the Afro-Arabian Rift System vol. 47. Rome: *Accademia Nazionale Dei Lincei,* p.275-292.

Kemp, J. (1982) *Reconnaissance Geology of the Harrat Lunayyir Quadrangle, 25/37 D. Saudi Arabian Department, Ministry for Mineral Resources Open File Report,* no. **BRGM-OF-02-18**.

Knott, S.T., Bunce, E.T., and Chase, R.L. (1966) Red Sea seismic reflection studies, the world rift system. *Can. Geol. Surv. Paper.* no. **66-14**, 33-61.

Kogbe, C.A. (1989) African rifting (editorial). *J. Afr. Earth Sci.* **8.**

Kohn, B.P., and Eyal, M. (1981) History of uplift of the crystalline basement of Sinai and its relations to opening of the Red Sea as revealed by fisson track dating of apatites. *Earth Planet. Sci. Lett.* **52**, 129.

Kuo, L.-C., and Essene, E.J. (1986) Petrology of spinel harzburgite xenoliths from the Kishb plateau, Saudi Arabia. *Contrib. Min. Petrol.* **93**, 335.

Labrecque, J.L., and Zitellini, N. (1985) Continuous sea-floor spreading in the Red Sea: An alternative interpretation of magnetic anomaly pattern. *Amer. Assoc. Petrol. Geol. Bull.* **69**, 5134.

Laughton, A.S. (ed.) (1966) The Gulf of Aden, in Relation to the Red Sea and the Afar Depression of Ethiopia. *Can. Geol. Surv. Paper.* no. 66-14.

Laughton, A.S. (1970) A new bathymetric chart of the Red Sea. *Phil. Trans. R. Soc. Ser A* **267**, 227.

Le Bas, M.J., Le Maitre, R.W., Streckeisen, A., and Zanettin, B. (1986) A chemical classification of volcanic rocks based on the total-alkali-silica diagram. *J. Petrol.* **27**, 745.

Le Pichon, X., and Cochran, J.R. (1988) The Gulf of Suez and Red Sea rifting: International workshop on the Gulf of Suez and Red Sea-rifting. *Tectonophysics* **153**, 320.

Le Pichon, X., and Francheteau, J. (1978) A plate tectonic analysis of the Red Sea–Gulf of Aden area. *Tectonophysics* **46**, 369.

Le Pichon, X., Francheteau, J., and Bonnin, J. (1973) *Plate Tectonics.* Amsterdam: Elsevier Scientific Publishing Co.

Lipparini, T. (1954) Contributi alla conoscenza geologica del Yemen (SW Arabia). *Bull. Serv. Geol. Ital.* **76**, 95.

Lister, G.S., Etheridge, M.A., and Symonds, P.A. (1991) Detachment models for the formation of passive continental margins. *Tectonics* **10**, 1038.

Lovelock, P.E.R. (1984) Review of the tectonics of the northern Middle East region. *Geol. Mag.* **121**, 577.

Lowell, J.D., and Genik, G.J. (1972) Seafloor spreading and structural evolution of the southern Red Sea. *Bull. Amer. Assoc. Petrol. Geol.* **56**, 247.

Lyberis, N. (1988) Tectonic evolution of the Gulf of Suez and Gulf of Aqaba. *Tectonophysics* **153**, 209.

Lyberis, N., Yurur, T., Chorowicz, J., Kasapoglu, E., and Gundogdu, N. (1992) The east Anatolian fault: oblique collisional belt. *Tectonophysics* **204**, 1.

MacFadyen, W.A. (1930) The geology of the Farasan Islands, Gizan, and Kamarin Island, Red Sea, Part I – General geology. *Geol. Mag.* **67**, 310.

MacFadyen, W.A. (1932) On the volcanic Zubayr Islands, Red Sea. *Geol. Mag.* **69**, 310.

Madden, C.T., Naqvi, I.M., Whitmore, F.C.Jr., Schmidt, D.L., Langston, W., Jr., and Wood, R.C. (1980) Paleocene vertebrates from coastal deposits in Harrat Hadan area, At Taif region, Kingdom of Saudi Arabia. *U.S. Geological Survey Open File Report* no. **80-0227**.

Madden, C.T., Schmidt, D.L., and Whitmore, F.C., Jr., (1983) Mastherium (Artiodactyula, Anthracotheriidae) from Wadi Sabya, southwestern Saudi Arabia: An earliest Miocene age for continental rift-valley volcanic deposits of the Red Sea margin. *U.S. Geological Survey Open File Report.* no. **USGS-OF-03-61**.

Makris, J. (1983a) Red Sea. The subcommission Deep Seismic Sounding of the European Seismological Commission (Activity report 1976-1980).

Makris, J. (1983b) Seismic investigations of the northern part of the Red Sea. *Dt. Forsch. Germ. Rep.*

Makris, J., and Rihm, R. (1987) Crustal stucture of the Red Sea region derived from seismic and gravity data. *Terra Cognita* **7**, 294.

Makris, J., and Rihm, R. (1991) Shear-controlled evolution of the Red Sea: pull apart model. *Tectonophysics* **198**, 441.

Makris, J., Menzel, H., Zimmerman, J., Bonjer, K.P., Fuchs, K., and Wohlenberg, J. (1970a) Crustal and upper mantle structure of the Ethiopian rift derived from seismic and gravity data. *Z. Geophys.* **36**, 387.

Makris, J., Thiele, P., and Zimmerman, J. (1970b) Crustal investigations from gravity measurements at the scarp of the Ethiopian Plateau. *Z. Geophys.* **36**, 299.

Makris, J., Menzel, H., and Zimmerman, J. (1972) A preliminary interpretation of the gravity field of Afar, northeast Ethiopia. *Tectonophysics* **15**, 31.

Makris, J., Menzel, H., Zimmerman, J., and Gouin, P. (1975) Gravity field and crustal structure of north Ethiopia. In: *Afar Depression of Ethiopia.* Bad Bergzabern, Schweizerbart, p.135.

Makris, J., Allam, B.A., and Moller, L. (1981) Deep seismic studies in Egypt and their interpretation. *EOS Trans. Amer. Geophys. Union* **62**, 230.

Makris, J., Henke, C.H., Egloff, F., and Akamaluk, T. (1991a) The gravity field of the Red Sea and East Africa. *Tectonophysics* **198**, 369.

Makris, J., Mohr, P. and Rihm, R. (1991b) Red Sea: Birth and early history of a new ocean basin. *Tectonophysics* **198**, 239.

Makris, J., Tsironidis, J., and Richter, H. (1991c) Heatflow density distribution in the Red Sea. *Tectonophysics* **198**, 383.

Manetti, P., Capaldi, G., Chiesa, S., Civetta, L., Conticelli, S., Gasparon, M., La Volpe, L., and Orsi, G. (1991) Magmatism of the eastern Red Sea margin in the northern part of Yemen from Oligocene to present. *Tectonophysics* **198**, 181.

Marinelli, G., Quia, R., and Santacroce, R. (1980) Volcanism and spreading in the northernmost segment of the Afar rift (Gulf of Zula). Rome: *Accademia. Nazionale Dei Lincei.* **47**, 421.

Mart, Y. (1982) Incipient spreading center in the Gulf of Elat, northern Red Sea. *Earth Planet. Sci. Lett.* **60**, 117.

Mart, Y., and Hall, J.K. (1984) Structural trends in the northern Red Sea. *J. Geophys. Res.* **89**, 11, 532.

Mart, Y., and Ross, D.A. (1987) Post-Miocene rifting and diapirism in the northern Red Sea. *Mar. Geol.* **74**, 173.

Martinez, F., and Cochran, J.R. (1988) Structure and tectonics of the northern Red Sea: catching a continental margin between rifting and drifting. *Tectonophysics* **150**, 1.

Martinez, F., and Cochran, J.R. (1989) Geothermal measurements in the northern Red Sea: Implications for lithospheric thermal structure and mode of extension during continental rifting. *J. Geophys. Res.* **94**, 12,239.

McCulloch, M.T., Gregory, R.T., Wasserburg, G.J., and Taylor, H.P. Jr. (1981) Sm–Nd, Rb–Sr, and $^{18}O/^{16}O$ isotopic systematics in an oceanic crustal section: evidence from the Samail ophiolite. *J. Geophys. Res.* **86**, 2721.

McGuire, A.V. (1987) Petrology of mantle and crustal inclusions in alkali basalts from western Saudi Arabia: Implications for formation of the Red Sea. Ph.D. Thesis, Stanford University.

McGuire, A.V. (1988a) Petrology of mantle xenoliths from Harrat al Kishab: the mantle beneath western Saudi Arabia. *J. Petrol.* **29**, 73.

McGuire, A.V. (1988b) The mantle beneath the Red Sea margin: xenoliths from western Saudi Arabia. *Tectonophysics* **150**, 101.

McGuire, A.V., and Bohannon, R.G. (1989) Timing of mantle upwelling: evidence for a passive origin for the Red Sea Rift. *J. Geophys. Res.* **94**, 1677.

McGuire, A.V., and Coleman, R.G. (1986) Petrology of the Jabal Tirf gabbro and associated rocks of the Tihama Asir complex, southwest Saudi Arabia and the opening of the Red Sea. *J. Geol.* **94**, 652.

McKenzie, D., Davies, D., and Molnar, P. (1970) Plate tectonics of the Red Sea and East Africa. *Nature* **226**, 243.

Mechie, J., and El-Esa, Z.H. (1988) Upper lithosphere deformations in the Jordan–Dead Sea transform regime. *Tectonophysics* **153**, 153.

Mechie, J., and Prodehl, C. (1988) Crustal and uppermost mantle structure beneath the Afro-Arabian rift system. *Tectonophysics* **153**, 103.

Menzies, M., Boscene, D., El-Nakhal, H.A., Al-Khirbash, S., Al-Kadasi, M.A., and Al-Subbary, A. (1990) Lithospheric extension and opening of the Red Sea: sediment–basalt relationships in Yemen. *Terra Nova* **2**, 340.

Merla, G. (1979) *A geologic map of Ethiopia and Somalia, scale 1:2,000,000.* Rome: Consiglio Naz. Reirche.

Milkereit, B. and Fluh, E.R., (1985) Saudi Arabian refraction profile: Crustal structure of the Red Sea-Arabian shield transition. *Tectonophysics* **111**, 283–298

Miller, A.R. (1964) Highest salinity in the world oceans? *Nature* **203**, 590.

Miller, A.R., Densmore, C.D., Degens, E.T., Hathaway, J.C., Manheim, F.T., McFarlin, P.F., Pocklington, R., and Jokela, A. (1964) Hot brines and recent iron deposits. *Geochem. Cosmochim. Acta* **30**, 341.

Miller, P.M., and Barakat, H. (1988) Geology of the Safaga concession, northern Red Sea, Egypt. *Tectonophysics* **153**, 123.

Miller, S.P., Macdonald, K.C., and Lonsdale, P. (1985) Near bottom magnetic profiles across the Red Sea. *Mar. Geophys. Res.* **1**, 401.

Mitchell, D.J.W., Allen, R.B., Salama, W., and Abouzakm, A. (1992) Tectonostratigraphic framework and hydrocarbon potential of the Red Sea. *J. Petrol. Geol.* **15**, 187.

Mittlefehldt, D.W. (1986) Petrology of high pressure clinopyroxenite series xenoliths, Mount Carmel, Israel. *Contrib. Min. Petrol.* **94**, 245.

Mohr, P. (1970) The Afar triple junction and sea-floor spreading. *J. Geophys. Res.* **75**, 7340.

Mohr, P. (1971) Ethiopian rift and plateaus: some volcanic petrochemical differences. *J. Geophys. Res.* **76**, 1967.

Mohr, P. (1978) Afar. *A. Rev. Earth Planet. Sci.* **6**, 145.

Mohr, P. (1982) Musings on continental rifts. In: G. Palmason (ed.) *Continental and Oceanic Rifts, vol. 8.* Washington, D.C.: American Geophysical Union, p. 293.

Mohr, P. (1983) Ethiopian flood basalt province. *Nature* **303**, 577.

Mohr, P. (1987) Structural style of continental rifting in Ethiopia: reverse decollments. *EOS Trans. Amer. Geophys. Union* **68**, 721.

Mohr, P. (1989) Nature of the crust under Afar: new igneous, not thinned continental. *Tectonophysics* **167**, 1.

Mohr, P. (1991) Structure of Yemeni Miocene dike swarms, and emplacement of coeval granite plutons. *Tectonophysics* **198**, 203.

Mohr, P., and Zanettin, B. (1989) The Ethiopian flood basalt province. In: T.D. Macdougall (ed.) *Continental Flood Basalts.* Dordrecht: Kluwer, p.63.

Moltzer, J.G., and Binda, P.L. (1981) Micropaleontology and palynology of the middle and upper members of the Shumaysi formation, Saudi Arabia. *Bull. Fac. Earth Sci., King Abdulaziz Univ.* **4**, 57.

Monin, A.S., Litvin, V.M., Podreazhansky, A.M., Sagalevich, A.M., Sorokhtin, O.G., Voitov, V.I., Yastrebov, V.S., and Zonenshain, L.P. (1982) Red Sea submersible research expedition. *Deep Sea Res.* **29**, 361.

Montenant, C., Ott d'Estevou, P., Purser, B., Burollet, P.F., Jarrige, J.-J., Orszag-Sperber, F., Philobbos, E., Plaziat, J.-C., Prat, P., Richert, J.-P., Roussel, N., and Thiriet, J.-P. (1988) Tectonic and sedimentary evolution of the Gulf of Suez and the northwestern Red Sea. *Tectonophysics* **153**, 161.

Montenant, C., Angelier, J., Beaudoin, B., Bolze, J., Burollet, P.F., Orzsag-Sperber, F., and Richert, J.-P. (1990) The western margin of the Red Sea, north of Port-Soudan. *Bull. Soc. Geol. France* **6**, 435.

Mooney, W.D., and Prodehl, C. (1984) Proceedings of the 1980 workshop of the International Association of Seismology and Physics of the Earth's Interior on the seismic modeling of laterally varying structures: Contributions based on data from the 1978 Saudi Arabian Refraction Profile. *U.S. Geol. Surv. Circ.* **937**, 158.

Mooney, W.D., Gettings, M.E., Blank, H.R.J., and Healy, J.H. (1985) Saudi Arabian seismic deep refraction profile: A traveltime interpretation of crustal and upper mantle structure. *Tectonophysics* **111**, 173.

Morgan, P., Swanberg, C.A., Boulos, F.K., Hennin, S.F., El-Sayed, A.A., and Basta, N.Z. (1980) Geothermal studies in northeast Africa. In: *Geol. Surv. Egypt.* **10**, 971.

Morton, W.H., and Black, R. (1975) Crustal attenuation in Afar. In: *Afar Depression of Ethiopia, vol. 1.* Stuttgart, Schweizerbart, p.55.

Mottl, M.J., and Holland, H.D. (1978) Chemical exchange during hydrothermal alteration of basalt by seawater – I. Experimental results for major and minor components of seawater. *Geochim. Cosmochim. Acta* **46**, 1103.

Myers, J.S. (1980) Structure of the coastal dyke swarm and associated plutonic intrusions of East Greenland. *Earth Planet. Sci. Lett.* **46**, 407.

Nakamura, N. (1974) Determination of REE, Ba, Fe, Mg, Na, and K in carbonaceous and ordinary chondrites. *Geochim. Cosmochim. Acta* **38**, 757.

Nasir, S. (1990) K-Ar age determinations and volcanological evolution of the northern part of the Arabian Plate, Jordan. *Eur. J. Min.* **2**, 188.

Nasir, S. (1992) Geochemistry and petrogenesis of Cenozoic volcanic rocks from the northwestern part of the Arabian continental alkali basalt province, Jordan. *Bull. Volcan.*, in press.

Nasir, S., and Al-Fuqha, H. (1988) Spinel-lherzolite xenoliths from the Aritain volcano, NE Jordan. *Min. Petrol.* **38**, 127.

Nawab, Z.A. (1984) Red Sea mining: a new era. *Deep Sea Res.* **31**, 813.

Neumann, A.C., and McGill, D.A. (1962) Circulation of the Red Sea in early summer. *Deep Sea Res.* **8**, 223.

Nicolas, A. (1985) Novel type of crust produced during continental rifting. *Nature* **315**, 112.

Nicolas, A., Boudier, F., and Montigny, R. (1987) Structure of Zabargad Island and early rifting of the Red Sea. *J. Geophys. Res.* **92**, 461.

Oberli, F., Htaflos, T., Meier, M., and Kurat, G. (1987) Emplacement age of the peridotites from Zabargad Island (Red Sea): a zircon U-Pb study. *Terra Cognita* **7**, 334.

O'Connor, T.E. (1992) The Red Sea–Gulf of Aden; hydrocarbon evaluation of multinational sedimentary basins. *J. Petrol. Geol.* **15**, 121.

Ohmoto, H., and Skinner, B.J. (1983) The Kurroko and related volcanogenic massive sulfide deposits: introduction and summary of new findings. In: H Ohmoto and B.J. Skinner (eds) *The Kuroko and Related Volcanogenic Massive Sulfide Deposits*. New Haven: Economic Geology, p.1.

Omar, G.I., Kohn, B.P., Lutz, T.M., and Faul, H. (1987) The cooling history of Silurian to Cretaceous alkaline ring complexes, southeastern Desert, Egypt, as revealed by fission-track analysis. *Earth Planet. Sci. Lett.* **83**, 94.

Orszag-Sperber, F., and Plaziat, J.C. (1990) The continental sedimentation (Oligo-Miocene) in depressions of the proto-rift of northwestern Red Sea (Egypt). *Bull. Soc. Geol. France* **6**, 385.

Ott d'Estevou, P., Jarrige, J.-J., Montenat, C., Prat, P., Richert, J.-P., and Thiriet, J.-P. (1987) Main structural aspects of the Gulf of Suez and the Red Sea rifting. In *Genese et Evolution des Bassins Sedimentaires, vol. 21*. Paris: Total Compagnie Francaise des Petroles, p.168.

Ottonello, G., Piccardo, G.B., Joron, J.L., and Treuil, M. (1980) Nature of the deep crust and uppermost mantle under the Assab region (Ethiopia): inferences from petrology and geochemistry of mafic-ultramafic inclusions. Rome: *Atti Dei Convegni Lincei.* **47**, 463.

Overstreet, W.C., Stoeser, D.B., Overstreet, E.F., and Goudarzi, G.H. (1977) Tertiary laterite of the As Sarat Mountains, Asir Province. *Saudi Arabian Dir. Gen. of Mineral Res. Bull.* **21**, 1.

Pallister, J.S. (1984) Reconnaissance geology of the Harrat Hutaymah quadrangle, Sheet 26/42 A, Kingdom of Saudi Arabia. U.S. Geological Survey.

Pallister, J.S. (1987) Magmatic history of Red Sea rifting: perspective from the central Saudi Arabian coastal plain. *Geol. Soc. Amer. Bull.* **98**, 475.

Pallister, J.S., Stacey, J.S., Fischer, L.B., and Premo, W.R. (1987) Arabian shield ophiolites and the late Protozoic microplate accretion. *Geology* **15**, 320.

Paul, S.K. (1990) People's Democratic Republic of Yemen: a future oil province, Classic petroleum provinces. *Geol. Soc. Lond.* **50**, 329.

Pautot, G. (1983) Les fosses de la mer Rouge: approche geomorphogique d'un stade initial d'ouverture oceanique realise a l'aide du Seabeam. *Oceanol. Acta* **6**, 235.

Pautot, G., Guennoc, P., Coutelle, A., and Lyberis, N. (1984) Discovery of a large brine deep in the northern Red Sea. *Nature* **310**, 133.

Perry, S.K., and Schamel, S. (1990) The role of low-angle normal faulting and isostatic response in the evolution of the Suez Rift, Egypt. *Tectonophysics* **174**, 159.

Petrini, R., Joron, J.L., Ottonello, G., Bonatti, E., and Seyler, M. (1988) Basaltic dykes from Zabargad Island, Red Sea: petrology and geochemistry. *Tectonophysics* **150**, 229.

Phillips, J.D. (1970) Magnetic anomalies in the Red Sea. *Phil. Trans. R. Soc. Ser. A* **267**, 205.

Phillips, J.D., and Ross, D.A. (1970) Continuous seismic reflexion profiles over the Red Sea. *Phil. Trans. R. Soc Ser. A* **267**, 143.

Picard, L. (1970) On Afro-Arabian graben tectonics. *Geol. Rdsch.* **59**, 337.

Piccardo, G.B., Messiga, B., and Vannucci, R. (1988) The Zabargad peridotite–pyroxenite association: petrological constraints on its evolution. *Tectonophysics* **150**, 135.

Piccirillo, E.M., Justin-Vistenin, E., Zanettin, B., Joron, J.L., and Treuil, M. (1979) Geodynamic evolution from plateau to rift: major and trace element geochemistry of the Central Eastern Ethiopian plateau volcanics. *N. Jb. Geol. Paleont. Abh.* **158**, 138.

Pilger, A., and Rosler, A. (1974a) *Afar Depression of Ethiopia, vol. 1.* Stuttgart: Schweizerbart.

Pilger, A., and Rosler, A. (1974b) *Afar Depression of Ethiopia vol. 2.* Stuttgart: Schweizerbart.

Pilger, A., and Rosler, A. (1976) The contemporaneous tectonic events of the Indian Ocean and neighbouring areas. *Sonderdr. Abh. Brauns. Wiss. Ges.* **27**, 67.

Plafker, G., Agar, R., Asker, A.H., and Hanif, M. (1987) Surface effects and tectonic setting of the 13 December 1982 North Yemen earthquake. *Bull. Seis. Soc. Amer.* **77**, 2018.

Plaumann, S. (1975) Some results of a detailed gravimetric survey of the southern Red Sea. *Geol. Jb.* **D13**, 155.

Plaziat, J., Purser, B.H., and Soliman, M. (1990) The geometry and dynamics of Miocene marine sediments and their relationship with the early tectonics of the northwest edge of the Red Sea rift. *Bull. Soc. Geol. France* **6**, 397.

Potorrf, R.J., and Barnes, H.L. (1983) Mineralogy, geochemistry, and ore genesis of hydrothermal sediments from the Atlantis II deep, Red Sea. *Econ. Geol. Monogr.* **5**, 198.

Powers, R.W., Ramirez, I.F., Redmond, C.D., and Elberg, E.L.J. (1966) Sedimentary geology of Saudi Arabia, geology of the Arabian Peninsula. *U.S. Geological Survey Prof. Paper* no. **560-D**, 1D-147D.

Prodehl, C. (1985) Interpretation of a seismic-refraction survey across the Arabian shield in western Saudi Arabia. *Tectonophysics* **111**, 247.

Prodehl, C., and Mechie, J. (1991) Crustal thinning in relationship to the evolution of the Afro-Arabian rift system: a review of seismic-refraction data. *Tectonophysics* **198**, 311.

Purser, B.H., and Hotzl, H. (1988) The sedimentary evolution of the Red Sea rift: a comparison of the northwest and northeast margins. *Tectonophysics* **153**, 193.

Purser, B.H., Orszag-Sperber, F., Plaziat, J.-C., Philobbas, E., Soliman, M., Montenant, C., Angelier, J., Bolze, J., Burollet, P.F., and Richert, J.-P. (1990a) Gulf of Suez and Red Sea. *Bull. Soc. Geol. France* **6**, 371.

Purser, B.H., Philobbas, E.R., and Soliman, M. (1990b) Sedimentation and rifting in the northwest parts of the Red Sea: a review. *Bull. Soc. Geol. France* **6**, 371.

Quennell, A.M. (1956) Tectonics of the Dead Sea rift. In: *Proceedings of the 20th Int. Geol. Congress, Mexico*, p.385.

Quennell, A.M. (1958) The structural and geomorphic evolution of the Dead Sea rift. *Q. J. Geol. Soc. Lond.* **114**, 1.

Quershi, I.R. (1971) Gravity measurements in the northeastern Sudan and crustal structure of the Red Sea. *Geophys. J. R. Astr. Soc.* **24**, 119.

Ressetar, T., Nairn, A.E.M., and Monrad, J.R. (1981) Two phases of Cretaceous–Tertiary magmatism in the eastern Desert of Egypt: paleomagnetic chemical and K-Ar evidence, Egypt. *Tectonophysics* **73**, 169.

Richter, H., Makris, J., and Rihm, R. (1991) Geophysical observations offshore Saudi Arabia: seismic and magnetic measurements. *Tectonophysics* **198**, 297.

Rihm, R., Makris, J., and Moller, L. (1991) Seismic surveys in the northern Red Sea: asymmetric crustal structure. *Tectonophysics* **198**, 279.

Roberston Research International (1988) *Geologic map of Sudan, scale 1:1,000,000.* Khartoum: Sudan Geological Research Authority.

Robson, D.A. (1971) The structure of the Gulf of Suez (Clysmic) rift, with special reference to the eastern side. *Q. J. Geol. Soc. Lond.* **115**, 247.

Roeser, H.A. (1975) A detailed magnetic survey of the southern Red Sea. *Geol. Jb.* **13D**, 131.

Romanowicz, B. (1991) Seismic tomography of the Earth's mantle. *A. Rev. Earth Planet. Sci.* **19**, 77.

Rona, P.A. (1980) *Seafloor Spreading Centers, Hydrothermal Systems.* Stroudsburg: Dowden, Hutchinson and Ross.

Rona, P.A. (1988) Hydrothermal mineralization at oceanic ridges. *Can. Miner.* **26**, 431.

Ross, D.A. (1972) Red Sea hot brine area; revisited. *Science* **175**, 1455.

Ross, D.A. (1983) The Red Sea. In: B.H-Ketchum (ed.) *Estuararies and Enclosed Seas.* Amsterdam: Elsevier, p.293.

Ross, D.A. and Degens, E.T. (1969) Shipboard collection and preservation of sediment samples collected during Chain 61 from the Red Sea. In: E.T. Degens and D.A. Ross (eds) *Hot Brines and Recent Heavy Metal Deposits in the Red Sea.* New York: Springer-Verlag, p.363.

Ross, D.A. and Schlee, J. (1973) Shallow structure and geologic development of the southern Red Sea. *Geol. Soc. Amer. Bull.* **84**, 3827.

Ross, D.A. and Schlee, J. (1977) Shallow structure and geologic development of the southern Red Sea. *Miner. Res. Bull. (Directorate General of Mineral Resources, Jiddah, Saudi Arabia)* **22**, E1.

Ross, D.A., Hays, E.E., and Allstrom, F.C. (1969) Bathymetry and continuous seismic profiles of the hot brine region of the Red Sea. In: E.T. Degens and D.A. Ross (eds) *Hot Brines and Recent Heavy Metal Deposits in the Red Sea.* New York: Springer-Verlag, p.82.

Ross, D.A., Whitmarsh, R.B., Ali, S.A., Boudreaux, J.E., Coleman, R.G., Fleisher, R.L., Girdler, R.W., Manheim, F., Matter, A., Ngrini, C., Stoffers, P., and Supko, P.R. (1973) Red Sea drillings. *Science* **179**, 377.

Ruegg, J.G. (1975) Main results about the crustal and upper mantle structure of the Djibouti region (T.F.A.I.). In: A. Pilger and A. Rosler (eds) *Afar Depression of Ethiopia.* Bad Bergzabern: Schweizerbart, p.120.

Ryan, M.P. (1987) Neutral buoyancy and the mechanical evolution of magmatic systems. In: *Magmatic Processes: Physochemical Principles. Geochemical Society Special Publication,* **1**, p.259.

Saffarini, G.A., Nasir, S., and Abed, A.M. (1985) A contribution to the petrology and geochemistry of the Quaternary–Neogene basalts of Central Jordan. *Dirasat* **12**, 138.

Said, R. (1962) *The Geology of Egypt.* New York: Elsevier.

Said, R. (1969) General stratigraphy of the adjacent land areas of the Red Sea. In: E.T. Degens and D.A. Ross (eds) *Hot Brines and Recent Heavy Metal Deposits in the Red Sea.* New York: Springer-Verlag, p.71.

Said, R. (1990) *The Geology of Egypt.* Brookfield, Vermont: A.A. Balkema.

Sardar, Z. (1978) Red Sea muds ripe for exploitation. *Nature* **275**, 582.

Savoyat, E., and Balcha, T. (1989) Petroleum exploration in the Ethiopian Red Sea. *J. Petrol. Geol.* **12**, 187.

Scheuch, J. (1976) Preliminary heat flow map of the Red Sea and an attempt to provide a geological–geophysical interpretation. In: A. Pilger and A. Rosler (eds) *Afar between Continental and Oceanic Rifting, vol. 2*. Stuttgart: Schweizerbart, p.171.

Schilling, J.-G. (1969) Red Sea floor origin: rare-earth evidence. *Science* **165**, 1357.

Schilling, J.-G. (1973) Afar mantle plume: rare earth evidence. *Nat. Phys. Sci.* **242**, 2.

Schilling, J.-G., Kingsley, R.H., Hanan, B.B., and McCully, B.L. (1992) Nd–Sr–Pb isotopic variations along the Gulf of Aden: Evidence for mantle plume–continental lithosphere interaction. *J. Geophys. Res.* **97**, 10,297.

Schmidt, D.L., and Hadley, D.G. (1985) Stratigraphy of the Miocene Baid formation, southern Red Sea coastal plain, Kingdom of Saudi Arabia. *U.S. Geological Survey Open File Report*, no. **85-241**.

Schmidt, D.L., Hadley, D.G., and Brown, G.F. (1983) Middle Tertiary continental rift and evolution of the Red Sea in southwestern Saudi Arabia. *U.S. Geological Survey Open File Report*, no. **83-0641**.

Schoell, M. (1976) Heating and convection within the Atlantis II deep geothermal system of the Red Sea. In: *Second United Nations Symposium on the Development and Use of Geothermal Resources, San Francisco, 1974*, vol. 1, p.583.

Schoell, M., and Hartmann, M. (1978) Changing hydrothermal activity in the Atlantis II deep geothermal system. *Nature* **274**, 784.

Schoell, M., Backer, H., and Baumann, A. (1974) *The Red Sea geothermal Systems, New Aspects on their Brines and Associated Sediments*. Proceedings of the International Association Genesis of Ore Deposits (IAGOD), Varna Meeting.

Searle, R.C., and Ross, D.A. (1975) A geophysical study of the Red Sea axial trough between 20.5° and 22° north. *Geophys. J.R. Astr. Soc.* **43**, 555.

Sebai, A., Feraud, G., Giannerini, G., Campredon, R., and Bertrand, H. (1987) ^{39}Ar-^{40}Ar dating on Cenozoic mafic volcanics of the Arabian plate associated with the early Red Sea opening. *Terra Cognita* **7**, 201.

Sebai, A., Zumbo, V., Feraud, G., Bertrand, H., Hussain, A.G., Giannerini, G., and Campredon, R. (1991) ^{40}Ar/^{39}Ar dating of alkaline and tholeiitic magmatism of Saudi Arabia related to the early Red Sea rifting. *Earth Planet. Sci. Lett.* **104**, 473.

Sellwood, B.W., and Netherwood, R.E. (1984) Facies evolution in the Gulf of Suez area: sedimentation history as an indicator of rift initiation and development. *Mod. Geol.* **9**, 43.

Sestini, J. (1965) Cenozoic stratigraphy and depositional history, Red Sea Coast, Sudan. *Bull. Amer. Assoc. Petrol. Geol.* **49**, 1453.

Seyler, M., and Bonatti, E. (1988) Petrology of a gneiss-amphibolite lower crustal unit from Zabargad Island, Red Sea. *Tectonophysics* **150**, 177.

Shanks, W.C. III, and Bischoff, J.L. (1980) Geochemistry, sulfur isotope composition and accumulation rates of the Red Sea geothermal deposits. *Econ. Geol.* **75**, 445.

Shimron, A.E. (1990) The Red Sea line – a late Proterozic transcurrent fault. *J. Afr. Earth Sci.* **11**, 95.

Shukri, N.M., and Basta, E.Z. (1954) Petrography of the alkaline volcanic rocks of Yemen, Egyptian University, scientific expedition of S.W. Arabia. *Inst. Desert Egypte Bull.* **36**, 130.

Siedner, G., and Horowitz, A. (1974) Radiometric ages of late Cainozoic basalts from northern Israel: chronostratigraphic implications. *Nature* **250**, 23.

Simkin, T., Siebert, L., McClelland, L., Bridge, D., Newhall, C., and Latter, J.H. (1981) *Volcanoes of the World*. Stroudsburg: Dowden, Hutchinson and Ross.

Spohner, R., and Oleman, P. (1986) *Topography, Red Sea Region (1:8,000,000).* Karlruhe: German Research Foundation.

Stacey, J.S., and Hedge, C.E. (1984) Geochronological and isotopic evidence for early Proterozoic crust in the eastern Arabian shield. *Geology* **12**, 310.

Steckler, M.S. (1981) The thermal and mechanical evolution of Atlantic-type continental margins. Ph.D. Thesis, Columbia University.

Steckler, M.S. (1985) Uplift and extension at the Gulf of Suez – indications of induced mantle convection. *Nature* **317**, 135.

Steckler, M.S., Berthelot, F., Lyberis, N. and Le Pichon, X. (1988) Subsidence in the Gulf of Suez: implications for rifting and plate kinematics. *Tectonophysics* **153**, 249.

Steen, G. (1982) Radiometric age dating and tectonic significance of some Gulf of Suez igneous rocks. In: *6th Petrol. Explor. Seminar, EGP, Cairo.*

Stein, M., Hofmann, A.W., and Goldstein, S.L. (1987) Lithospheric evolution of the northern Arabian shield: Nd and Sr isotopic evidence from basalts, xenoliths and granites. *Terra Cognita* **7**, 418.

Steinitz, G., Bartov, Y., and Hunziker, J.C. (1978) K-Ar age determination of some Miocene–Pliocene basalts in Israel: their significance to the tectonics of the Rift Valley. *Geol. Mag.* **115**, 329.

Stern, R.J., Gottfried, D., and Hedge, C.E. (1984) Late Precambrian rifting and crustal evolution in the northeastern desert of Egypt. *Geology* **12**, 168.

Stoffers, P. and Kuhn, R. (1974) Red Sea evaporites: a petrographic and geochemical study. *Initial Repts. Deep Sea Drilling Project* **23**, 849.

Stoffers, P., and Ross, D.A. (1977) Sedimentary history of the Red Sea. *Miner. Res. Bull.* **22**, D1 (*Saudi Arabia, Directorate General of Mineral Resources, Jiddah, Saudi Arabia*).

Stosser, D., and Camp, V.E. (1985) Pan-African microplate accretion of the Arabian Shield. *Geol. Soc. Amer. Bull.* **96**, 817.

Styles, P., and Gerdes, K.D. (1983) St. John's Island Red Sea: a new geophysical model and its implications for the emplacement of ultramafic rocks in fracture zones and continental margins. *Earth Planet. Sci. Lett.* **65**, 353.

Suayah, I., Rogers, J.J.W., and Dabbagh, E. (1991) High-Ti continental tholeiites from the Aznam trough, northwestern Saudi Arabia: evidence of "abortive" rifting in the "embryonic" stage of Red Sea opening. *Tectonophysics* **191**, 75.

Sudan (1963) *Sudan geological map, scale 1:4,000,000 (3rd edn.).* Khartoum: Sudan Survey Department.

Swartz, D.H., and Arden, D.D.J. (1960) Geologic history of the Red Sea area. *Amer. Assoc. Petrol. Geol. Bull.* **44**, 1621.

Tanimoto, T., and Anderson, D.L. (1985) Lateral heterogeneity and azimuthal anisotropy of the upper mantle: Love and Rayleigh waves 100–250 s. *J. Geophys. Res.* **90**, 1842.

Tard, F., Masse, P., Walgenwitz, F., and Gruneisen, P. (1991) The volcanic passive margin in the vicinity of Aden, Yemen. *Bull. Centres. Rech. Explor. Prod. Elf-Aquitaine* **15**, 1.

Tazieff, H., Varet, J., Barberi, F., and Giglia, G. (1972) Tectonic significance of the Afar (or Danakil) depression. *Nature* **235**, 144.

Thrope, R.S., and Smith, K. (1974) Distribution of Cenozoic volcanism in Africa. *Earth Planet. Sci. Lett.* **22**, 91.

Tierclelin, J.J., Taieb, M., and Faure, H. (1980) Continental sedimentary basins and volcano-tectonic evolution of the Afar rift. In: *Geodynamic Evolution of the Afro-Arabian rift system.* Rome: *Atti Dei Convegni Lincei.* **47**, 491.

Tramontini, C., and Davies, D. (1969) A seismic refraction survey in the Red Sea. *J. R. Astr. Soc.* **17**, 225.

Tromp, S.W. (1950) The age and origin of the Red Sea graben. *Geol. Mag.* **87**, 385.

Uchupi, E., and Ross, D.A. (1986) The tectonic style of the northern Red Sea. *Geol. Mar. Lett.* **5**, 203.

U.S.G.S. (1963) *Geologic map of the Arabian Peninsula.* U.S.G.S. Misc. Geol. Inves. Map I-270-A, 1:2,000,000. Washington, D.C.: U.S. Geological Survey.

Vail, J.R. (1978) *Geologic Map of Sudan, scale 1:2,000,000,* Inst. Geol. Sci. Overseas Geol. Miner. Res. no. 49. London: HMSO.

Vail, J.R. (1985) Pan-African (Late Precambrian) tectonic terrains and the reconstruction of the Arabian–Nubian shield. *Geology* **13**, 839.

Vail, J.R. (1988) *Lexicon of Geologic Terms for the Sudan.* Brookfield: A.A. Balkema.

Van den Boom, G. (1968) Zur Petrogenese der Plateaubasalte Nordostjordaniens. *Geol. Jb.* **85**, 489.

Vaslet, D. (1990) Upper Ordovician glacial deposits in Saudi Arabia. *Episodes* **13**, 147.

Vening Meinesz, F.A. (1934) Gravity expedition at sea II, 1923–1932. Delft: *Netherlands Geod. Commun. Publ.* vol. 2.

Verzhbitskiy, Y.V. (1980) Studies of the heat flow in the rift zone of the Red Sea. *Oceanology* **20**, 580.

Villa, I.M. (1988) $^{40}Ar/^{39}Ar$ analysis of amphiboles from Zabargad Island (Red Sea) (abstract). *Tectonophysics* **150**, 249.

Voggenreiter, W., and Hotzl, H., (1989) Kinematic evolution of the southwestern Arabian continental margin: implications for the origin of the Red Sea. *J. Afr Earth Sci.* **8**, 541.

Voggenreiter, W., Hotzl, H., and Jado, A.R. (1988a) Red Sea related history of extension and magmatism in the Jizan area (southwest Saudi Arabia); indication for simple-shear during early Red Sea rifting. *Geol. Rdsch.* **77**, 257.

Voggenreiter, W., Hotzl, H., and Mechie, J. (1988b) Low-angle detachment origin for the Red Sea Rift system? *Tectonophysics* **150**, 51.

Wegener, E. (1924) *The Origin of Continents and Oceans.* London: Methuen.

Wernicke, B. (1985) Uniform-sense normal simple-shear of the continental lithosphere. *Can. J. Earth Sci.* **22**, 108.

White, R.S., and McKenzie, D. (1989) Magmatism at rift zones: The generation of volcanic continental margins and flood basalts. *J. Geophys. Res.* **94**, 7685.

White, R.S., Spence, G.D., Fowler, S.R., McKenzie, D., Westbrook, G.K., and Bowen, A. N. (1987) Magmatism at rifted continental margins. *Nature* **330**, 439.

Whiteman, A.J. (1968) Formation of the Red Sea depression. *Geol. Mag.* **105**, 231.

Whiteman, A.J. (1971) *The Geology of the Sudan Republic.* Oxford: Clarendon Press.

Whitmarsh, R.B., Weser, O.E., Ross, D.A., and others. (1974) *Initial reports of the Deep Sea Drilling Project (Red Sea, v. 23).* Washington, D.C.: U.S. Government Printing Office.

Wilson, M. (1989) *Igneous Petrogenesis.* Boston: Unwin Hyman.

Worzel, J.L. (1965) Pendulum gravity measurements at sea 1936–1959. *Lamont. Geol. Obs. Contrib.* **807**.

Young, R.A., and Ross, D.A. (1970) Volcanic and sedimentary processes in the Red Sea trough. *Deep Sea Res.* **21**, 289.

Yousif, I.A. (1982) On the structure and evolution of the Red Sea. Ph.D. Thesis, University of Louis Pasteur, Strasbourg.

Zanettin, B. (1980) Introduction to the symposium. In: *Geodynamic Evolution of the Afro-Arabian Rift System.* Rome: *Atti Dei Convegni Lincei.* **47**, 21.

Zanettin, B., Justin-Visentin, E., Nicoletti, M., and Piccirillo, E.M. (1980a) Correlation among Ethiopian volcanic formations with special references to the chronological and stratigraphical problems of the "Trap Series". In: *Geodynamic Evolution of the Afro-Arabian Rift System.* Rome: *Atti Dei Convegni Lincei.* **47**, 231.

Zanettin, B., Justin-Visentin, E., and Piccirillo, E.M. (1980b) Migration of the Ethiopian continental rifts in the course of the Tertiary evolution of the Afro-Arabian rift system. In: *Geodynamic Evolution of the Afro-Arabian Rift System.* Rome: *Atti Dei Convegni Lincei.* **47**, 251.

Zhang, Y.-S. and Tanimoto, T. (1992) Ridges, hotspots and their interaction as observed in seismic velocity maps. *Nature* **355**, 45.

Zierenberg, R.A. (1990) Deposition of metalliferous sediment beneath a brine pool in the Atlantis II Deep, Red Sea. In: G.R. Murray (ed.) *Gorda Ridge. A Seafloor Spreading Center in the United States' Exclusive Economic Zone.* New York: Springer-Verlag, p.131.

Zierenberg, R.A., and Shanks, W.C., III (1986) Isotopic constraints on the origin of the Atlantis II, Suakin and Valdiva brines, Red Sea. *Geochim. Cosmochim. Acta* **50**, 2205.

Zierenberg, R.A., and Shanks, W.C., III (1988) Isotopic studies of epignetic features in metalliferous sediment, Atlantis II deep, Red Sea. *Can. Miner.* **26**, 737.

Zindler, A., and Hart, S. (1986) Chemical geodynamics. *A. Rev. Earth Planet. Sci.* **14**, 493.

Zonenshain, L.P., Monin, A.S., and Sorokhtin, O.G. (1981) Tectonics of the Red Sea rift near 18 degrees north. *Geotektonika* **2**, 3 (in Russian).

Index